CAMBRIDGE LIBRARY COLLECTION

*Books of enduring scholarly value*

## Botany and Horticulture

Until the nineteenth century, the investigation of natural phenomena, plants and animals was considered either the preserve of elite scholars or a pastime for the leisured upper classes. As increasing academic rigour and systematisation was brought to the study of 'natural history', its subdisciplines were adopted into university curricula, and learned societies (such as the Royal Horticultural Society, founded in 1804) were established to support research in these areas. A related development was strong enthusiasm for exotic garden plants, which resulted in plant collecting expeditions to every corner of the globe, sometimes with tragic consequences. This series includes accounts of some of those expeditions, detailed reference works on the flora of different regions, and practical advice for amateur and professional gardeners.

## The Art and Practice of Landscape Gardening

In the late nineteenth century, British garden design was dominated by two opposing schools, those of 'landscape' and of 'the formal garden'. Henry Ernest Milner (1845–1906) was the son of Edward Milner, a practitioner in the landscape tradition who was a colleague of Sir Joseph Paxton, but Henry did not begin working with his father until after a career as a civil engineer which took him to North America and Russia. In this 1890 book, a number of the exemplar gardens are taken from Milner's father's designs, with two studies of his own work (one in England and one in Hungary) at the end, and he surveys the whole 'art and practice', from aesthetic theory to building and planting, with plans and illustrations. Reginald Blomfield's 1892 *The Formal Garden in England* (also reissued in this series) was written partly as a response to this work.

Cambridge University Press has long been a pioneer in the reissuing of out-of-print titles from its own backlist, producing digital reprints of books that are still sought after by scholars and students but could not be reprinted economically using traditional technology. The Cambridge Library Collection extends this activity to a wider range of books which are still of importance to researchers and professionals, either for the source material they contain, or as landmarks in the history of their academic discipline.

Drawing from the world-renowned collections in the Cambridge University Library and other partner libraries, and guided by the advice of experts in each subject area, Cambridge University Press is using state-of-the-art scanning machines in its own Printing House to capture the content of each book selected for inclusion. The files are processed to give a consistently clear, crisp image, and the books finished to the high quality standard for which the Press is recognised around the world. The latest print-on-demand technology ensures that the books will remain available indefinitely, and that orders for single or multiple copies can quickly be supplied.

The Cambridge Library Collection brings back to life books of enduring scholarly value (including out-of-copyright works originally issued by other publishers) across a wide range of disciplines in the humanities and social sciences and in science and technology.

# The Art and Practice of Landscape Gardening

H.E. Milner

CAMBRIDGE
UNIVERSITY PRESS

University Printing House, Cambridge, CB2 8BS, United Kingdom

Cambridge University Press is part of the University of Cambridge.
It furthers the University's mission by disseminating knowledge in the pursuit of education, learning and research at the highest international levels of excellence.

www.cambridge.org
Information on this title: www.cambridge.org/9781108076401



This edition first published 1890
This digitally printed version 2016

ISBN 978-1-108-07640-1 Paperback



The original edition of this book contains a number of colour plates, which have been reproduced in black and white. Colour versions of these images can be found online at www.cambridge.org/9781108076401

# THE ART AND PRACTICE OF
# LANDSCAPE GARDENING.

THE

# ART AND PRACTICE

OF

# LANDSCAPE GARDENING.

BY

HENRY ERNEST MILNER, F.L.S., Assoc. M. Inst. C.E.

---

*WITH PLANS AND ILLUSTRATIONS.*

London:

OF THE AUTHOR, DULWICH WOOD, NORWOOD, S.E.,

AND

SIMPKIN, MARSHALL, HAMILTON, KENT, AND CO., LIMITED.

STATIONERS' HALL COURT, LONDON.

1890.

CHARLES DICKENS AND EVANS,
CRYSTAL PALACE PRESS.

# PREFACE.

---

My father, the late Edward Milner, as the colleague for many years of Sir Joseph Paxton, was concerned in all the later achievements of Landscape Gardening carried out by that distinguished man; and from 1850 until 1884 himself designed and completed many of the finest works of the kind that have ever been produced, not only in this country, but in various notable places on the continent of Europe. By this prosecution of his art in such extended practice, he attained a purely exceptional experience, the opportunity for which ripened his artistic powers; and without question he was enabled to illustrate by his works a steady advance of the art, which is essentially English. On my part, as the colleague of my father, and as successor not only to his profession but to many fruits of his experience, I too have had ample opportunities to practically illustrate the art that I love and the work that I delight in. Impelled by that love, I have endeavoured to realise the principles on which I have been led to base the artistic conceptions of my work, and the points of practice by which it may be carried out. I do not intend to advance any fashion, past or present, in Landscape Gardening, but to set forth my own opinions and my own practice, however incompletely that may be done.

H. E. M.

DULWICH WOOD, S.E.,
*June*, 1890.

# CONTENTS.

Warren Hine, R.I. pinxt

J. R. Hutchinson, sculpt.

# THE ART AND PRACTICE OF

# LANDSCAPE GARDENING.

---

## INTRODUCTION.

To define the properties of an art must always be an opiniative endeavour, since conceptions of beauty are varied and ever varying, and the means whereby art can express what is beautiful, appeal—with their fascination or their fuller force of conviction—rather in the measure of the recipient's appreciation than of the giver's power.

The function of fine art is to exhibit beauty—that ineffable beauty which can stir in the human breast emotions claiming for our better nature kindred with higher things. The architect rears a temple or a cathedral in which effects of light and the combination of lines create a feeling of awe that compels an ignorant and careless man to speak in a whisper; the sculptor or the painter records in form, or simulated form and colour, that conception of beauty with which the artist has been gifted; the composer gives being, by musical tone, to some subtle emotion of our nature that has yearned dumbly within us, till the touch of divine genius gives it distinctness of existence. The poet makes beauty articulate to our reason, seeking that path to the heart. But there is another form of art yet. There are expressions of beauty to which we are all subject, and to an appreciation of which we are most of us educated. We are always under their influence. In a thousand forms Nature herself puts forward her charms, and the influence of these is, in degree, upon us all, arousing emotions, or enriching those we possess. It is difficult to define the working and result of such emotions; moreover, the wealth of their sweetness and grandeur would be lessened could we impose on ourselves a definite conception of their extent. Nature appeals to us in multitudinous ways, in particular as well as in general effect: in the mute appeal of the violet, as well as in the majesty of the oak; in light and shade; in the sheen of water; in the infinite variety of the formation of the earth's surface; in the disposition of foliage both in colour and outline; in combination of different features, and in constant development. Simply to accumulate and crowd together natural productions, in themselves beautiful, may give us collections, but not the desired result. If we endeavour to define the art of landscape gardening, it may be stated as the taking true cognizance of Nature's means for the expression of beauty, and so disposing those means artistically as to co-operate for our delight in given conditions.

It is not the author's intention by this book to give another history of landscape gardening; his intention is rather to record the present development of the art and his own relation to, and practice of it. The love of natural picturesqueness is of very modern growth in England. Its rise was coincident with that of the modern novel in literature, and of *genre* painting in art. Little more than a century has, with the aid of conducing influences, ripened the feeling to a passion. Peace and augmenting prosperity have been necessary conditions for advancing the art of gardening in this country, since that time when safety made it possible for the wealthy to live outside towns in unfortified houses. Up to the middle of the last century all progress made was in fetters of artificiality, formal and imitative in respect of the art features. The garden contiguous to the house was laid out for utility or for pleasure with a purpose of distinguishing it from the surrounding scenery; and it was natural that the same feeling which gave the character to the house should also prevail in the garden. Every possible appearance of uniformity was bestowed on the enclosed ground. Regularity and order contrasted with Nature's aspect outside. Admiration was courted by means of regularity, and by marking to the spectator the labour, design, and expense bestowed on the garden. Moreover, in any extension, duplication of lines was aimed at rather than the introduction of a new design. But, as the eighteenth century came on, the newer influences were strengthening; causes produced effects, which in their turn became causes of advance, and with the appreciation of Nature's beauties in scenery, the art that was to minister to the better feeling became more definite in its tendency, until the present development of the Art of Landscape Gardening—truthfully and distinctively styled English—has been reached. Our love of natural beauty has been greatly fostered by the modern systems of easy travelling, by peace at home and by the stupendous accretion of private as well as of national wealth. It is startling to note in how narrow a strip of our history, these conditions have been established completely. The Victorian period has these facts on its records.

Up to the commencement of the eighteenth century, landscape gardening was mostly the work of architects; and it was characterised by formal art features, by intricacy of design in parts, and by treatment of the ground as a plane surface. That ideal of beauty resulted in a feeling that Nature had been pushed, and coerced, and sometimes dragged to an effect. Nature, it is true, is always at work, and, as time goes on, clothes a neglected grave with beauty, or makes an unsightly rubbish heap lovely with her gifts; and, in like manner, she gives venerableness to magnificently-grown trees in stately parks, which compel admiration not entirely due to the design of the planter. So natural beauty combines with historical associations to give many old gardens an indescribable charm especially valued in this land. Next, the influence of pictorial artists was felt. They strove to modify the old constrained practice by the imposition of an almost equal artificiality of picturesqueness. The resulting effect, as now appreciated, at all events, is a feeling of unrest and stiffness. Nature seems still to work in fetters. I shall endeavour now briefly to investigate some of the conclusions which, in my estimation, form bases for a better practice of the art, the method of which I have sought to describe in this book.

Water, by its constant gentle action, or by its sweeping directed force, has, without question, been a principal agent in forming the natural surface of our earth. It has scooped out valleys and modified the hills; in one place leaving their rounded sides and tops covered with forest, in another bringing down loose earth till the hill-tops and slopes show clear and perfectly curved outlines; or, again, leaving bare the rocks standing out in grand abruptness. With the subsidence of the water, wide grassy valleys have been formed; wherein are seen long vistas of lawn running up till they are lost in the obscurity of the forest. Then, through the middle of the hollowed and widening fertile valley, runs the stream, that spreads into the lake, giving an expression of the water's subsidence in the restricted action of the stream or the subsided force of the more placid water. Therefore it is right, in forming a restricted landscape, to bear in mind Nature's grand agent in the formation of her greater landscapes, wherein are presented endless varieties of form, of line, and of colour, far-stretching even plains rich with herbage, hill-sides and tops covered with foliage, as well as swelling downs with their vast sweeping curves. These forms of beauty may exercise their fascination on the most unlearned, or appeal with augmented force to the most cultivated human nature. It is the province of the Landscape Gardener, as I understand the art, to appreciate the multitudinous means whereby Nature expresses her beauty, and so to use those means artistically as to arrange their force for producing the delightful result he desires to achieve. To servilely copy Nature's forms is to incur the pettiness of mere reproduction in little, with the penalty of a falsification. To utilise her means and to let the spirit of her works influence our art in every practicable way, is the true practice of the art of Landscape Gardening.

And can we deduce principles of art from the effect of Nature's picturesqueness, and the details of her action? A loving student of her, with an artistic spirit, does derive some principles; but he feels more than he can describe of them. He comes to know the truth that he who feels best the infinitude of Nature's expressions of beauty, receives their influence without consciousness of the process. By observation of mankind, he notes how certain mental processes become active when reason and the finer sympathetic spirit, that we call feeling, are pleasantly excited by testimony of the eye. He classifies such phenomena and calls them laws. The artist, after all, can do little more than generalise the truths in this wise he thinks he has discovered; yet they are the means by which he works, as with a gentle and practised method he designs to bring the sweet influences to bear, with as little evidence as may be of the shaping hand. For illustration, I shall very briefly indicate some such deductions that seem to me to afford guidance for the artist of Landscape Gardening.

Nature seldom presents a straight line in any of her forms, unless in the seeming regularity of an oceanic horizon, or the smaller line of water surface. A straight line is the product of art, for even the apparently upright line of the Parthenon columns results from a delicate curve. Nature presents in her broad effects and graduated detail, an infinity of curvilinear features. There is beauty in contrast of form; but appropriateness of natural position is a condition of its value in the scene. As in Nature, water plays such an important part in the formation of the land, so in our circumscribed landscape, the feeling that pervades all Nature should prevail in our treatment of the ground. A calculated shadow on a lawn is a resource of value for the artistic use of natural effect. The extended surface of down land shows exquisite gradation of light and shade; but not the true vastness of space to the unpractised vision. The eye seeks to estimate distant features, and insensibly gains a standard of measurement from intervening objects, and, when these are absent, most frequently miscalculates distance. Lines or objects placed in a direction going from the line of vision, make the space so marked appear longer, whilst lines running across it, make the space appear less distant. Grass, clothing the ground surface, has an expression of stability and repose; it seems to be immovable; and in colour it illustrates the tint of foliage to which it forms a base and background. Trees and shrubs, by the contrast of foliage, give variety, and a gradation of colours may promote the idea of distance. They should clothe the hill-tops and slopes in masses of irregular outline. A sky-line of trees should not be continuous, but should be broken. A valley appears deeper by not being planted, as a hill appears higher than it really is by being planted to its summit. Single trees emphasize falling ground, and they, like the shadowy regions of a wood, conduce to a sensation of mystery, subtly stimulating imagination. They induce an idea of possible shelter that bestows pleasurable sensations. Falling ground appears shorter, whilst level ground at the base of a hill, as also rising ground, seems longer than it in reality is. The idea of spaciousness can be artificially promoted, particularly by the breaking of continuous lines and hard boundary lines, and by providing various objects for the eye to count, just outside the direct line of sight. Vision invariably travels down a hollow, or depression, or through any opening. Thus the idea of distance may be created, and the eye be conducted to realise what is desired. By directing the vision to distant beauties of landscape, they may be brought into the artist's plan. Trees especially serve to frame a particular view. In every situation a beyond implies discovery, and affects the imagination. The area is circumscribed of which we can take cognizance too readily and completely; imagination is then confused or frustrated. The beauty of water, in motion or still, is of universal acceptance. The created character of a water-feature must be consonant with the surrounding land; for fitness to surrounding conditions is a measure of beauty to both. A lake expresses spaciousness; but much of its charm is due to its outline. A river expresses action. Trees or high banks on the edge of water diminish its extent when seen from the opposite bank, and make it dull. An opening in the trees, or the lowering of a high bank, make a gleam of light, and the length in that direction appears greater. The ground immediately around a dwelling, forming the artificial base on which it rests, should be treated formally, and the site and aspect of the house will, in a measure, determine the relation to it of the contiguous ground. The approaches should be direct, convenient, and not strained.

Such deductions as the foregoing may serve to illustrate some of the resources of the Landscape Gardener, and by their paucity in relation to the infinite developments of beauty in the nature around us, demonstrate the impossibility of defining the range of this art. Its conceptions are more frequently to be felt than described; and the attempt to deduce from them a set of fixed principles must be as futile as an endeavour to completely systematise the ever varying yet constant manifestations of beauty that we encounter in the vast field of Nature's operations which adorn the surface of our country. Nature is the great exemplar that I follow. The method I adopt I have endeavoured to describe in the several divisions of this book.

A

# THE APPROACH.

THE approach through an enclosed estate to the house upon it, is one of the most pressing matters to settle in laying out a residential property, and this feature often determines the choice of a site for the structure. The principal points for consideration are, firstly, the place for entrance from the public road, and next the route thence, requiring not only artistic treatment but convenience of approach, and of access to the house, the offices, the stables, kitchen-garden and the farm, and care that on the route such good views, or objects of interest that may exist, shall be displayed.

It may be taken as an axiom, that the approach to a house should always appear to be direct, and any deviation from such directness should not only arise from, but should also be made to appear to arise from, some decided obstacle. By direct, is not meant straight. A curved line of road is generally to be preferred, because it is more easy of construction, and because more varied views can be obtained, since it can follow in great measure the natural contour of the ground. A straight drive should be used only when an imposing, or somewhat pretentious building is at the end of it. It is only allowable in flat country, or where some special object has to be attained. If the ground is very undulating a straight road is out of character with its surrounding. Of course, when a venerable and stately avenue of old trees has to be dealt with, a different treatment is taken. Curves should be long and easy. The side slopes of a curved drive can be more easily and freely dealt with than the sides of a straight drive, because they may be steep or flat, as desired. An uniform slope is unnatural. It is advisable in deep cuttings to bring forward prominent points which may support planting; or, if in a suitable country, rocky promontories may be made or shown; and between these projections the spaces should be judiciously excavated, so that the vision following up the hollows thus made will be carried to the ground beyond, and the observer will not realise the fact that an artificial cutting has been introduced.

The natural slope of earth surface varies; but, as a rule, it is not wise to make any slope steeper than 2 : 1, save at exceptional points, where there is planting or rockwork. At such points of abrupt declivity the turf should, if necessary, be pegged down. The gradient of a curved drive can be varied, following within limits the natural undulation of the ground. It should not be made with a series of gradients of certain ratio meeting at angles; but the gradient should rise or fall imperceptibly, with a line as even, continuous, and graceful as that of a horizontal drive. At the entrance from the public road, and also where the route reaches the house, the gradient should be nearly level, and the line straight, and it is well to keep the gradient level at crossings, or turns to various points. The gradient of a good drive should not be steeper than one in fourteen, though one in nine is a road over which a carriage may be driven with safety. A drive should not run parallel to the public road with the mere purpose of lengthening the course, or seeming to prolong it. When the house is at a comparatively short distance from the public road, on a much higher level, and the object is apparent to overcome the difference of level, then the resource may be adopted. It is to be remarked that a curved line looks much straighter on paper than when worked out on the ground. If there is a continuous gradient on a continuous curve, with a dip in the gradient beyond which the drive is again seen, that part of the drive in the hollow should be either straight, or decidedly curved; otherwise, when seen from either end, the line appears to be broken.

A straight approach requires very careful treatment. It is artificial in character, and requires, as well as admits of, artificial expression. In undulating or on falling ground it should, if possible, be made at right angles to the slope of the hill. The gradient should be very even, and much flatter than that used for a curved line. Unless it is very even, although a graceful gradient curve be used, the observer looking up such a drive will imagine it is not straight. The slopes, both where cut and where artificially constructed by filling, are better if they are formal and regular; not with alternate cutting and filling as on a railway, but with uniformity of resource.

The width of drives is ruled by the character and importance of the traffic they are to take. A breadth of nine feet suffices for the passage of one carriage; where two carriages may meet, the road should be fourteen feet wide. The above is a minimum dimension, such as may be resorted to when the ground is flat, and on either side of the road is grass, to which a foot passenger can retire, or on to which even a carriage could in case of need be turned; but in ordinary cases it is better to give eleven or twelve feet for the single drive, and sixteen to eighteen feet for the double road. Economy sometimes rules in regard to these dimensions. The width of drives is frequently determined not by the exigencies of the traffic, but by the relatively important character of the route. Thus a drive to the principal entrance of the house would be fourteen feet, or sixteen feet or eighteen feet, while that to the stables or offices ten feet. Or as at Beechy Lees the drives are as shown on the sketch I. A drive with a sharp curve must have a greater width than is necessary in the straight.

As any one enters on an estate from the public road, he always looks around when just inside the gate, and perhaps, unconsciously, records an impression. It should be a matter of careful study, then, to give to the entrance as much realisation and promise of beauty as may be feasible. The difference between a dusty, untidy road outside, and the shaded, well-trimmed drive within the gate, should be made pleasurably apparent; and, as far as practicable, a view of the outlying grounds, a stretch of park, or the inner foliage of a wood should be visible through an opening, or be seen in broad expanse for the moment as the visitor passes rapidly on. At all cross roads and turning points, the same inspection occurs, and of necessity the opportunity is most favourable when the visitor alights at the entrance to the house, and contemplates the scene—the near view, or the distant outspread prospect—that he may afterwards look at so often, but never with more zest than when he receives, perhaps hardly noting it, his first impression.

Drives that leave the main route for unimportant points, should be curved as soon as may be convenient, so that the eye may not be tempted to explore them, and give to them a certain importance to the detriment of the main drive. It is preferable that the drive rise gradually to the house throughout its whole course; but if that be not practicable, it is important that it should have an ascending gradient at least during the last stretch, when the front entrance is in view. If the natural surface at this point be higher than the ground level of the house, it should be lowered so that there may be a rise, however slight, for however short a distance, at this termination of the road. Until the drive be made thus to rise, a straight direction should not be adopted, and a direct view

A
B
C
D
E
H
16
I
10
9
12
J

of the house should be prevented by planting, or by embankment, so that the idea that the house is in a low position may not be aroused. The drive should not skirt the garden, or overlook it. It is sometimes difficult to comply with the former condition; but the latter can generally be met by sinking the carriage way, by raising a bank along the gardens, and by planting. At Chatsworth, though a straight drive, which is much used by the public, skirts the west garden front, yet it is completely shut off from the gardens by a high wall retaining the terrace gardens.

The treatment of either end of the main carriage drive is of much importance. Taking first that nearest the house. The direction and level of the approach, and the character of the architectural features, will rule greatly the plan that can be adopted for the facilities of carriage traffic, which must be provided here. For most ordinary purposes the turn (*see Plan*) is sufficient, but examples of different methods are shown. The gravelled plain in front of the porch should not be less than thirty-three feet for a small house; but forty to sixty feet are requisite where two or more carriages may stand. It is well at places of importance to arrange this end of the drive, so that waiting carriages may circulate. There must be a sufficient space on each side of the porch, and in line with it, to admit of carriages drawing up close to the door, and to progress beyond it without too abruptly turning. The entrance to the gardens should not be from the drive, but from the house, as, for example, in the turn to a small house. (*See Plan.*)

In choosing the place for, and forming, an entrance from the public road (about the artistic effect of which something has already been said) advantage may be taken of a turn in the highway, if the position be suitable, so that by adapting the line of the public road, and continuing it to the entrance gates, much importance may be given them by making it appear that the highway leads through them. (*See Plan, fig. A*, and sketches, p. 9.) The entrance gates may be formed at the junction of two roads as at *B*; or where a cross road—even a minor one—comes on to the main road as at *C*, or the entrance can be set back as *D*; or as *E*, where the gates are set sufficiently back from the public road to allow a carriage to stand clear. To a straight drive the entrance should be imposing; and though any of the forementioned examples may be adopted, yet it is well that the treatment of this work should be formal in character. (*See Plan, fig. H.*) The gates as well as the lodge should be at right angles to the drive, and belong to it rather than to the public road. The line of boundary outside the gates should also be straight for, at any rate, a few feet, before any turn is begun.

The lodge should be so placed that the windows in the living rooms command a certain length of both the drive and the public road.

In placing such an entrance to the estate it is well, if possible, not to make it at the boundary of the property. One consideration in fixing the position should be the direction of the principal traffic likely to pass to it; the position of the town, the village, the church, of notable places or objects of interest. It is moreover advisable to plan the entrance, if it can be, at the foot of a hill or rise in the public road, and not part way up an ascent, or at the top of it. A drive that goes off fairly level ground at the foot of a hill, always promotes a feeling of more repose than is to be experienced in the positions just mentioned.

The actual work of making drives is dealt with later under the heading of General Formation.

# SITE OF THE HOUSE AND OF THE GARDENS.

---

THERE are certain leading considerations that should rule our determination in choosing the site for either a house or a garden, which considerations may almost be taken as independent of taste, and more in the nature of necessary conditions. They may be defined as (*a*) aspect—relation to the points of the compass; (*b*) prospect—relation to the surrounding view; (*c*) natural shelter; (*d*) convenience in regard to the approaches, and in respect of communication with stables and out-offices, as well as with the gardens; (*e*) the levels of the land; (*f*) the formation of the subsoil.

Each of these questions will be treated in turn.

(*a*).—The forms of houses, and the position of the principal rooms in them, are so varied that it is difficult to put down a rule that shall be absolute for all, even in the matter of aspect. In the southern and midland counties of England, however, there is no doubt that the main line of the house should be S.E. If the sides of a house form a square, and the front be to the S.E., it would, as the sun in summer rises N. of E., and sets N. of W., have sunlight on all its walls; whereas if the front of the house were placed due S., the N. face would lack sunshine, while that to the S. would probably have too much. The effect of such conditions on the comfort of the house is enormous and constant, and should never be undervalued. A house thus placed enjoys favourable conditions that are important, but are not very readily recognised. The sun exercises its fullest force between 1 and 2 p.m. At that hour the rays fall at an angle on the wall, and so with an indirect impact that saves much oppressive heat, without undue loss of illumination; in fact, the extremes of heat and cold are modified. In the northern counties, it may be advisable to give the house a more southern aspect when it is a consideration to gain full advantage of the shorter period of sunshine, and particularly when a series of the best chambers have to be brought under the influence.

(*b*).—It is of course an important consideration that such beauty as may be derived from a prospect should be, so far as practicable, open to the dwellers in the best apartments of the house, and the position of the apartments may often be ruled accordingly. But it is not desirable that the aspect of a residence should be sacrificed to secure such a prospect. If need be, other positions may be created in the gardens or grounds wherefrom the beauties of the view may be fittingly contemplated; but the conditions of the aspect are fixed in their relation. No doubt all these questions are matters of compromise in the end, but the steps whereby such compromise is arrived at must be very warily and advisedly taken. If it be determined that a fine view may be taken better from a position in the garden than from the windows, another point of attractiveness is created. Indeed, in country places especially, it is a gain to create particular points of interest, such as a fine prospect, a tennis-ground, a rose garden, an old-fashioned herbaceous garden with trimmed hedges, or walls covered with climbers, a (so-called) American garden, a fernery, a lake, etc. Such objects, beautiful and pleasurable in themselves, yet requiring a slight excursion to reach them, are less commonplace than the stables, or even the hothouses, to which the regulated trips are often made.

Even a view, however beautiful, that is lacking in variety of prominent features, becomes tame to the constant onlooker, though to him who occasionally contemplates it, the effect is fresh and delicious. When the prospect is simply over a long stretch of even country, the eye may get tired of it, and welcome the growth of intervening trees that serve to break up the too regular panorama. When the view is of a mountainous district, or where the surface is broken by abrupt hills, its ever-changing colours, its light and shade, under sun and cloud, destroy monotony. This is the character of extended scenery to be sought for. Water is always an agreeable feature, especially when there is some promontory or curved shore to diversify the regularity of margin.

In the circumscribed area of a park, or a garden, where the boundary forms our limit of view, there are modes of treatment that may modify or remove the impression of contracted space. The groups of trees with undergrowth (technically called "planting"), or even single trees, by the colour of the foliage, and by the disposing of the boundary plantations, can promote the idea of spaciousness. Such plantations should not exhibit continuous lines, unless for some strong reason, such as the creation of necessary shelter. It is sometimes better even to let in the view of the neighbouring town or fields. When any object in itself disagreeable, or by association unpleasant, lies within view of the house, or of any point near the house that has been made notable, care must be taken to block out the object by an intermediate group of planting; in short, a wall, even though it be of foliage, by too regular form, marks an enclosure too directly, and militates against the idea of spaciousness and freedom, where spaciousness and freedom should be most fully expressed. Any hard formal boundary line, however wide its range may be, is inimical to proper effect in this sense.

When we contemplate any landscape, the vision invariably travels down a hollow or depression, natural or artificial, and the eye seeks to estimate the most distant features first, then to gain from intervening objects its measure, or presumed measure, of distances. Thus the idea of distance can be created and artistically adapted. In arranging the position of its principal chambers in a house, it will be well to consider these facts when deciding on the site, so that, so far as may be practicable, the rooms may be well appropriate to the available views, or points of vantage in that respect, existing, or to be created. The eye may be, as it were, led to realise certain beauties, under the conditions; but the possibility of such conditions must be considered in choosing the *site.* Sometimes the existence of a natural feature — a group of old trees, or even a single tree—may go far toward determining many questions of the choice. A venerable tree is a feature in decorative work, having its fifty or sixty years of priceless value, that must not be ignored or underestimated, because it may, and should be, connected with the base of the house, and in such degree with the house itself, to which, and to the whole design, it brings by its mere age a characteristic appropriateness of adornment that is of the highest value. It may be that a site has to be chosen in this densely-populated land, where the least objectionable points of view have to be considered rather than natural beauties of wide prospect. In such circumstances, the site of the residence should be kept as low as may be, and by raising the surrounding ground, and by planting, all that is offensive may be shut out. With every choice a house is well placed on a southern hillside, preferably on a slight spur, at about one-third distance from the top, so that shelter may be obtained from the hill, and by planting above the house.

(*c*).—The question of natural shelter has an obvious connection with the particular district selected, and with the winds that prevail there. The most natural and best protection is found in a wooded hill. Wind force is more effectually modified and tempered by growing trees than by any other means. If no hill exist, or the hill be quite bare, planting should be employed. It should extend, and may be liberally used, on north-eastern, northern, and north-western sides. In some districts, notably on the southern coasts, such shelter is to be obtained by planting the south-western side. Dr. Herbert Watney is of opinion that it is not conducive to health to have a mass of foliage close to the house on its south-western side; but he lays down an opposite rule in regard to the north-eastern side. It is frequently evident to any one who stands at the south-western corner of a wood that the north-eastern wind is tempered by its passage through the trees, whereas the contrary position, with a south-westerly wind passing through the trees, gives the watcher a damp and disagreeable experience. Trees break the force of the wind, in such circumstances, better than does a brick wall, apart even from the question of picturesqueness. Trees break and disperse the waves of wind; walls divert the wind into draughts.

(*d*).—Convenience of approach to a house, and the relation of its position to high roads and lines for public travelling to necessary, or valuable, stations, to town or village, and to the distances from all these, are considerations of serious importance. We may, of course, make our main drive with proper directness to the house, and well-planned branching communications with the stables and offices, through the park or estate; but the position of all these in regard to the surrounding conditions of public approaches, must be an element of action in choice of site, and in modifying the details of it. More is said on this subject under the head of THE APPROACH.

(*e*).—The natural levels of the land, however it may be possible to modify them by art, have a strong bearing on the selection of a site. The ground level of the house should always be placed high enough, and yet nothing should be done that may give an appearance that the structure is "perched up." This is an initial consideration ; and very frequently an error in judgment of this kind vitiates much of the gratification that is gained by the excellence of work in other details. The mistake is most frequently made when dealing with sloping and low ground, and in those circumstances should be most guarded against. We should carefully bear in mind that a house will never look perched up if sufficient ground base be given to it. The proper effect is gained by an arrangement of the contiguous surrounding ground-surface ; the relation of the house to the terrace (which is, in fact, the plateau of the building, with its graduating outlying features), and an artistic treatment of these lines in careful avoidance of abruptness. It is conservative of healthy conditions, as well as more tasteful, to keep the house, as it were, out of the ground ; and should the design not include terraced walks, or building constructions connected with them, yet the effect may, in degree, be obtained by earth-working. Soil may be taken from the upper side of the house site, or where the ground requires to be lowered, or where undulations have had to be created, and thus the material requisite may be obtained. The ground level of Lord Wolverton's house at Iwerne Minster, the site of which was of necessity fixed on level ground, was raised ten feet above the surrounding plain ; then by the making of a terrace garden, on the southern and eastern sides, and by graduating the raised ground between the house entrance on the northern side and the level, the desired effect was created, for it is not possible to detect that the house has not been built on a knoll. In a case where the ground rises steeply, or abruptly, on one side of the house, and it would be an unduly expensive work to take away, or utilise by transference, great quantities of the earth, deep valleys should be excavated up the steep banks, in proper places, with calculated irregularity of direction, and the promontories so created, should be planted with judgment. In suitable geological positions rocks may be developed, or rockwork used in like manner. Such hollows fix in the mind the general ratio of the slope. The base or ground-line of the house should always be level. It frequently occurs, however, that the ground-line of the offices is at a lower level. In such case the difference should be marked by low planting.

(*f*).—With regard to the subsoil, most frequently there is little room for choice. Much prejudice exists against building a residence on any but a gravelly or a sandy soil. Where a dwelling-house has to be erected on a fresh position, a porous soil is to be preferred; but in a fully inhabited district, where perfect drainage is carried out, and the whole site of the structure can be overlaid with a flooring of concrete, there need not be any fear of building on clay. In gravelly districts that have been formerly old river bottoms, especially when near to existing great rivers, or when they are low-lying, and in similar districts that are thickly inhabited, it is necessary to adopt precautionary means for preventing the exhalation of noxious gases through the light soil, from depths where percolated moisture may be stagnant; whereas in a moderately dry clay, although there may be a dampness that can be removed by efficient drainage, there will not be any harmful vapours.

The site of a house, the design and arrangement of the structure, the position of its chief entrance and of its principal apartments, greatly influence judgment as to the position and arrangement of the gardens. In due course the details of such arrangements will be more fully treated of.

If we imagine a place that possesses most desirable features, the site of a dwelling-house should have fine prospects to the south-east, and to the south-west. The principal approach and entrance should be on the north-western face, the offices on the north-eastern side, the stables and the kitchen garden beyond. The pleasure gardens should be on the south-eastern aspect, with a continuation towards the east; the south-western face might be open to the park. As one does not exhibit a beautiful vase on the floor, but on a proportionate pedestal, so the house should be made to appear to rest on some base that may dignify it. That effect is best achieved by giving straight lines of walk, or slope, wall, or balustrade, according with the levels of the ground, and with the architectural character of the house. Such base is called a terrace, and the space enclosed by the horizontal lines, a terrace garden, described in detail hereafter. The terrace garden is an artificial creation, and should show in every detail the hand of man, differing in this from the garden proper, which, though fine in calculated detail of its plan, should express by its breadth of treatment, but unmistakably, that nature has triumphed over art, because art has subtly tutored the development of nature's overwhelming beauty.

# THE TERRACE.

---

THE terrace is not only the narrow strip of raised level ground placed parallel with the house, or the more stately portion — often with architectural adornments — that is laid out along the face of the structure, but must be understood as the whole of the ground that forms the base, or setting, for the building.

The greatest divergence between the work of English and foreign landscape gardeners is to be seen in their several methods of dealing with the ground immediately surrounding the house. In England we lay down as an axiom that the treatment of ground next the house shall be artistically formal, with regular lines of turf, slopes, walks, or beds, all displaying harmony, so far as may be, with the architectural character of the building. On the Continent they surround the house with broad, irregularly curved spaces, or walks, that have nothing in common with the design of the structure. By one practice the endeavour is to give a base to the building, and to create on the contiguous ground an expression of kindred artistic spirit; by the other the ground is treated as something apart, and a feeling of unrest is created.

A terrace may have various forms, from the simple walk parallel with the house, to the more elaborate arrangement shown in the Plan. The larger, more important and decorated the building, the more extensive, massive, and ornate may be the terrace. It should have a definite proportion to the size of the house.

That part nearest the house should be level with its horizontal ground-line. Length of terrace gives importance to a plan. It has been laid down that the width of the terrace should be equal to the height of the front of the house, and the rule is right when applied to many grand buildings; but for a less pretentious dwelling-house, which is frequently high in comparison with its length, the proportion is not fitting. For such a building the width of the grass might be 10 ft.; that of the walk, 9 to 12 ft.; and of the grass thence to the edge of the slope, not less than 6 ft. If beds for flowers are to be cut on the flat space of grass, the width must necessarily be greater. The greater the depth of slope, the greater should be the distance between the edge of the walk and the slope-edge. To shorten this distance is one of the commonest mistakes made; it creates a sensation of falling off, or insecurity, that is inimical to comfort. When, however, a dwarf wall or balustrade is used, the space that would have been laid in grass in this position should be given to increase the width of the gravel. It is well to give a slight fall of 1 in. in 10 ft. in the space between the house and the edge of the slope. In determining the distance that the slope-edge, or wall, shall be from the building, care must be taken to avoid cutting off the view beyond. If a portion of the landscape be thus sacrificed, it may be necessary to form a slope quite close to the house, and then make the terrace walk on a lower level beyond. In any case, there should always be a level space at the top and at the bottom of the slope, which should extend its length in unbroken line, not following the irregularity of line that may be formed by the building at its earth level. If a break of any kind must be made in this continuous line, only one such interruption should be suffered. The effect of this slope next the house is to add to its apparent altitude, as the eye in-

C
D
E
G

sensibly estimates the height from the line of gravel below. The space between the edge of a terrace walk and the upper edge of its outer slope may not be less than 6 ft., but that between the bottom of a slope and the edge of a walk may be as little as 1 ft. 6 in. Generally slopes should be made in the ratio of 2⅓ to 1, if above 1 ft. 6 in. in depth. In large places, or where importance is required, and the depth is not less than 3 ft., the slope should be 2½ to 1; smaller slopes, such as for sunk panels, etc., will not have a ratio of more than 1½ to 1. Terrace slopes should not have a greater depth than 5 to 6 ft. Where a greater depth is needed, two slopes should be made with a level width between them of not less than 4 ft.

The arrangement of parterres in the terrace garden should always be formal. It will generally be found that there is some detail in the house itself, or connected with the period of its architecture, that may be used to characterise the arrangement of walks or beds, apart from the strict design of terrace walls, steps, or balustrades. The terrace garden is viewed chiefly from the house, where the spectator's eye is 5 ft. above the floor-line; therefore, in designing the beds for flowers or foliage plants, such beds should not be placed too near the house; they should be even in width, with as few acute angles as possible; longer than wide, as seen from the house. If surrounded by gravel, the spaces of bed and of gravel should be evenly disposed. Besides bedding plants, evergreen dwarf foliage plants may be largely introduced, and precise designs of clipped box and yew are not out of place, if the building has a character that is consonant with such accompaniment. Of foliage plants, those only should be selected that are dwarf in habit and grow evenly. The whole arrangement in this position is artificial, and this is a place where the recurrence to old fashions in gardening is desirable. Good work of this kind is chiefly shown by attention to minute details. In contemplating such a formal arrangement from the house, there are some noticeable effects: the semicircle —a form frequently adopted—appears flatter than it really is; lines from the spectator give an expression of length; cross-lines shorten the area in appearance. Sunk panels, if adopted, should not be deeper than 1 ft. 6 in., and the beds in them should not be near the edge at bottom of the slope.

Stone steps, used in connection with terrace slopes, should be made as plain and as solid in appearance as possible. For a 2⅓ to 1 slope the tread of such would be 14 in., the rise 6 in.; for a 2½ to 1 slope the tread would be 15 in., with a rise of 6 in. The best steps are made of solid Portland, or other good stone, built as shown in Plan II. At the Crystal Palace the steps are of granite. When the lateral extent of the step is too great to admit of one stretch of stone being used, the joints should be made to recur alternately, and sub-walls should be constructed to support the stones at each juncture; and the walls at each end should be built wide enough to carry the plinth, or balustrade that contains the whole flight, as well as the end of each tread. Moreover, there must not be any joint in the plinth at the angles, but it should appear in the slope. The nose of the topmost step should be in line with the upper edge of the slope. The steps and the plinth should range with the slope. The width of the plinth will vary to accord with the dimensions of the whole flight, but the ordinary rule is 1½ in. for each foot in the breadth of the flight. For any ordinary flight of terrace steps, the plinth should be from 3 to 6 in. above the line of the slope, and consequently of the steps. Vases chosen to ornament the structure should be imposed on square pedestals situated on the top and at the bottom of the

II

plinth. In selecting such vases, in which flowers are to be planted, choose those forms that admit of the flowers being placed close to the edge. The margin of the vase should rather tend inwards than outwards. It should not form an overhanging lip. In ordinary circumstances the plinth should be designed to harmonise with the architectural style of the house. When the flight is not connected with a terrace wall, or it stands apart from the slope, there should not be any balustrade, and ornamentation should be very sparse, or absent.

When walls are used in connection with terrace gardens, with or without balustrading, whether they take the place of the slope or are formed in conjunction with it, they should agree in architectural character with the house; and not only that, but be formed of the like material, so that they may seem closely a part of the general design. Such walls should never be less than 3 ft. in height or more than 10 ft., exclusive of the balustrade, which usually is made 3 ft. high. But there is an exception to this ruling. On flat ground, for the simple purpose of giving a base-line, or a prominent demarcation of the formal gardening from the natural, a simple plinth, 12 to 18 in. high, and of equal width, may be used with a certain effect, as, for example, at Gosforth House. But walls considered and used simply as retaining walls, not seen from below, and on which the turf rests, need not follow the architectural character of the house. The position of any terrace wall or balustrading must be very carefully considered, for it is inadmissible that it should interfere with the view of falling ground beyond it. It may be advisable to adopt a slope whose lower edge touches a flat parallel space coinciding with and supported by the top of a retaining wall. There should always be a level space at foot of every slope or wall.

There are two most general forms of terrace walls. One is parallel with the terrace walk and the line of the house, for which it seems to constitute a base; this may or may not be surmounted by a balustrade. From this terrace there may be steps conducting to a lower level, perhaps to the pleasure grounds, but possibly to the second terrace with its formal gardens, whence flights of steps give access to the naturally-treated landscape gardens. To this second terrace there is generally a balustrade, which bears the expression of a boundary. In either case the treatment is architectural. It is well to remember that an access other than by steps from the terrace to the gardens should be devised for the possible use of invalids.

As to dimensions, the following may be generally stated: vertical walls should have a thickness in cut stone of ·35 of the height; in brick, 4 of the height; in dry rubble, ·5 of the height. A wall with a "battered" or sloped front will need less material, and have strength equal to a vertical wall, both having an equal thickness of base. Care must be exercised, especially in falling ground, that the foundations are carried deep enough to prevent sliding. There must be drainage from the back of walls, either by pipes built through the structure at proper intervals of distance, or by the insertion of a porous backing from which pipes to carry off water communicate. A backing of sods, laid by hand, will prevent thrust, and admit of a lessened thickness in the wall. The back of the wall should be left rough in construction. The average slope of ground is 1½ ft. horizontal to 1 ft. vertical. The weight of dressed granite is 156 lbs. per cubic foot; sandstone 137 lbs., lime concrete 119 lbs., cement concrete 137 lbs., brickwork 112 lbs., sand 100 lbs., and of clay 119 lbs.

Molesworth gives the following formulæ for retaining walls:

E = Weight of Earthwork per cube yard.
W = Weight of Wall.
H = Height of Wall.
T = Thickness of Wall at top.
T = H × Tabular No.

| | E : W : : 4 : 5. | | E : W : : 1 : 1. | |
|---|---|---|---|---|
| BATTER OF WALL. | CLAY. | SAND. | CLAY. | SAND. |
| 1 in 4 . . | ·083 | ·029 | ·115 | ·054 |
| 1 in 5 . . | ·122 | ·065 | ·155 | ·092 |
| 1 in 6 . . | ·149 | ·092 | ·183 | ·118 |
| 1 in 8 . . | ·184 | ·125 | ·218 | ·153 |
| 1 in 12 . . | ·221 | ·160 | ·256 | ·189 |
| Vertical . . | ·300 | ·239 | ·336 | ·267 |

In calculating the strength of surcharged walls, substitute Y for H, Y being the perpendicular at the end of a line (L = H) measured along the slope to be retained.

Y = 1·71 H in slopes of 1 to 1.
= 1·55 H " 1½ to 1.
1·45 H " 2 to 1.
1·31 H " 3 to 1.
1·24 H " 4 to 1.

An etching is given of the terrace garden at Friar Park, Henley, the residence of Frank Crisp, Esq. It serves to illustrate what is meant by a base being made to a house. In this case walks lead from the forecourt, from the garden entrance, and from the conservatory, on to an upper terrace, from which walks descend to a lower terrace garden with its parterre, which, in its turn, gives place to the flowing lines of the pleasure gardens. In this case a low moulded plinth is fixed on the top of the grass slope.

# ARRANGEMENT AND FORMATION.

THE facts and considerations in this chapter are given in relation to the general plan of a garden. The several matters spoken of, as *Planting*, etc., are treated in detail elsewhere in the book.

For arranging the plan of a garden, few rules can be laid down, but artistic principles must be insisted on. So many considerations press in to vary design, that arbitrary dealing by imposition of what may be termed paper designs, however ingenious, is, in its degree, ill-advised. The detailed plan should spring from the site, as an adaptation of its natural or created natural features, and should not be, as it were, forced upon the position, crushing it to an artificial scheme. To copy simply the design of another place is inadmissible. Considerations that rule in this connection are almost infinite: extent, geological formation, soil, existing natural formation or features, climate and aspect, the display of distant beauty, conformity to outside influences, particularly to the requirements of the possessor, and the expenditure of money that may be made. It is this important variety of modifying influences, and how they are dealt with, that gives charm to each new work of landscape gardening, and to the developments it presents; just as we contemplate a fresh work of the kind in pictorial art, and note how the artist has treated the natural features, the colours and tints, and their modifying juxtaposition on the canvas. The painter, however, may indeed have his rules as to composition, for the use of his colours and the production of his distances. His picture is not a servile copy of Nature in its exact details, but an artistic rendering of the *effect* of Nature, as seen by his educated eye, and recorded by his skilful hand. His picture is always viewed from the same point; but in the nature-pictures created by the landscape gardener the point of view is on every side; there is no back to the canvas. In each position the object should be one of beauty, of interest, and of delight, and its relation to other features, and to the whole field of the spectator's vision, be closely and truly considered. The landscape gardener must remember that his colours change and grow; he must realise, as he creates his pictures, that in a few years what now seems like a light green stroke of pigment to the painter—but a complementary effect—will have become a tall tree, beautiful in itself, but of altered beauty, either helping or marring the landscape. He follows Nature by adapting or garnering her beauties, and tutoring her, so to speak, to the display of them; but by following Nature is not meant a slavish imitation or reproduction of any of her particular scenes. Some are unattractive, some very inappropriate, all are subject to dissimilar conditions; and imitation in Nature, as well as art, produces pettiness. But the spirit of the beauty of Nature, embodied, as it were, in those of her works or features that express her majesty, simplicity, peacefulness, sweetness, refinement, strength, and variety, in form, colour, abundance, or any of her multifarious aspects of loveliness, should be included and brought into juxtaposition in an ideal scene, so far as we are able to promote its natural development. And always be it remembered,

> Nature is made better by no mean,
> But nature makes that mean.

In forming the surface of the ground, as already pointed out, we should realise the fact that most of the graceful undulations of fertile soil, result from the action of water. Here is a natural agency the effect of which we may imitate, in modifying the surface in our circumscribed landscape; so that our treatment of the ground may be consonant with what is in reality a pervading expression. In Nature, straight lines are very rarely found; straight lines are the production of art. The terrace, as before explained, and the region immediately next the house, and indeed, in a less degree, that next any minor building in the grounds, being by position artificial, should be treated in a formal manner. This postulate will so far influence our treatment of the rest of the landscape, that as the house is approached from the boundaries of the estate, the planting, design and work should become finer, more intricate, and more careful. Beyond the line of the terrace wall, slope, or walk, the design of the garden may be naturally treated. In this position a feeling of freedom should assert itself, and it may be subtly induced by the undulating ground, with the curved lines of walk, and planting. From the terrace, walks should not go off in the same direction, but should, at any rate for some distance, deviate. In the general arrangement of such walks, the curves should be set out with broad sweeping lines, the chord of which should be so great, that each sweep should be hidden from the succeeding bend. A multiplication of meaningless walks should be avoided, as should be the creation of anything like the wriggling serpentine lines so often seen in badly designed villa-gardens. It should be generally apparent that each walk serves some special object, as the route to some distant point, to connect two lawns, to approach some particular group of planting, to reach some point of view, or to provide alternative means of communication at different levels. No doubt the designer will have first placed the group of planting, the lawn, the access to the view, or the variation in level, but such a fact must never be obtrusive, and the art which serves to emphasize the idea that the walk can only go rightly in one place, must not be apparent.

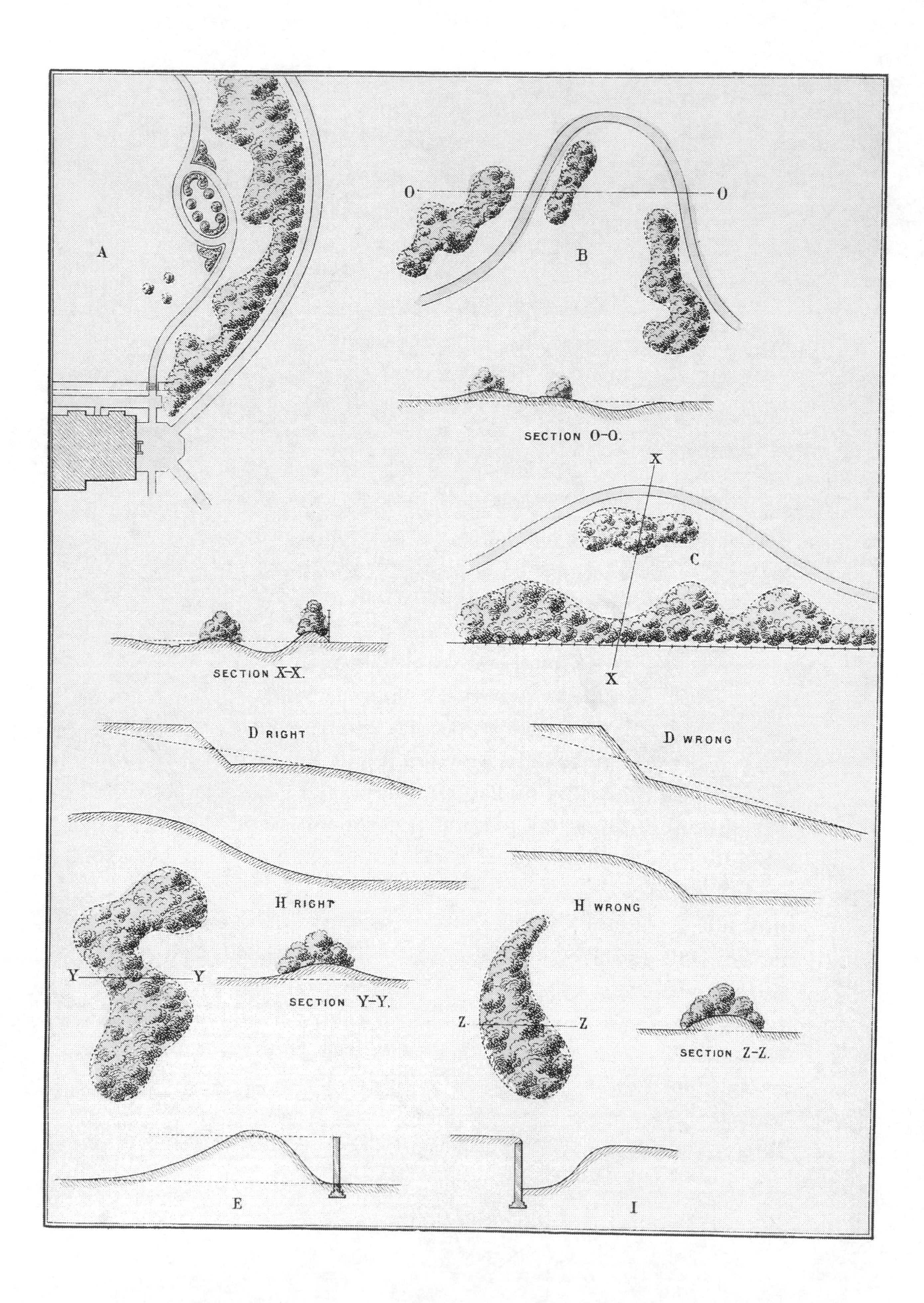
A
O
O
B
SECTION O-O.
X
C
SECTION X-X.
X
D RIGHT
D WRONG
H RIGHT
H WRONG
Y
Y
SECTION Y-Y.
Z
Z
SECTION Z-Z.
E
I

It is essential that groups of planting should be, and should appear to be, rightly placed in the general composition. If there be old single trees, or groups of trees, on the turf, they should be treated as specimens, and isolated each on its own mound, and not have little bushes put about them. There is an expression of affording shelter, whether from heat or rain, about such objects, and in that way they bestow a pleasing idea, while they contribute their beauty to the scene. In planting, variety of outline in the height is almost as desirable as variety of outline in the plan. A frequent error in laying out gardens is to make all the groups of planting assimilate, and all walks and undulations alike in appearance and outline. That is to be avoided. The proper method is to make the prominent points high, and to keep the intervening bays low, both in forming the ground, and in selecting the plants. In setting out groups of planting, uniformity of curve, and parts of circles should be avoided. Long bays with the turf running up them should be made; the spectator's vision is unconsciously led up them if only for the moment. As seen from the house, or its terrace, a plantation, which for example (*see Plan, fig. A*) may be one skirting the garden, screening it from observation on the approach, or protecting it from harmful winds, may be made very effective by marking the outline of the turf with long bays running up beyond the points, so that the eye cannot follow the whole outline. An appearance of much greater extent and freedom is attained. Variety in such designs is essential, and bald uniformity of parts is to be shunned. There is a killing tameness in such repetition of parts in the outline as is involved in the recurrent use of circles or parts of circles, or of egg-shaped figures such as are frequently to be seen by lines of walk in Continental gardens. These dicta apply equally to the height of trees, the character of the foliage and the tints, and especially to the undulations of the made ground, where uniformity and repetition destroy one of the most exquisite expressions of Nature's beauty—the wave line of the ground—by rendering it unnatural in form, burlesque in fact. Variety within limit of the natural means, and of such aids as art may supply, should indeed be a predominating aim. To that end to create fresh picturesqueness, to open out fresh views, and by contrasts of colour, of line, and of level to tempt Nature herself to exercise her fascinations, so direct on our conceptions, so difficult to bring into dry definition, is worthy of our best effort. The creation in well-chosen position of objects of special interest, such as a rose garden, an herbaceous enclosed garden, an old-fashioned allée of clipped yew, box, or lime, or a secluded rockery, filled with choice Alpine plants, through which, breaking by a cascade from the upper ground, a little stream rushes, and then meanders to the lake below; or a seat covered with climbing plants, that perhaps half hide some quaint inscription, or mark the site of a local legend of old time; these are all legitimate means to the same end. But of more consideration than all of them should be the general appearance of the surrounding ground, which must compel the conviction that here Nature is but tutored, not fettered, yet tempted to the expression of her beauties in the freedom of favouring conditions.

When the lines of distant views are determined, it is necessary that details be carefully considered, so that no breaks be introduced in the direct line of sight. In regard to one such view, walks may be arranged to lead in the direction of the best point for realising the prospect, and they may curve round this point, sunk below it; but the lawn up to that position must be hollowed out, or concave, so that the eye may be tempted to range up this valley, in the centre of which, or in the line of sight, there must not be planting, although there may be planting on either side (*see Plan, fig. B*); or, in another case, there may be a long stretch of lawn; in a third, especially if the view be narrowed to a single object, such as a church spire, a simple break in the planting, through which the vision will instinctively pierce, will be enough, and the object will appear, as it were, framed by the foreground foliage. If a curve occur in a boundary walk, an effective treatment is to make a small group of planting in the centre of the curve, and form a hollow in the ground between this group and the boundary plantation (*as fig. C*). The lowering of the walk may be so made that its depth will provide sufficient earth for raising the group, as well as the mound next the boundary. As a matter of fact, it is seldom necessary to bring earth from a distance for such works in the formation of a garden. In forming the terrace, when the ground recedes, the surface may be lowered by increasing the depth of the slope till sufficient earth is provided for the requisite filling (*as fig. D*). To raise mounds for planting, hollowed out pathways in the curves may be made between groups; a whole area, as well as the walks, may be lowered when it becomes necessary to secure soil for an adjacent planting. Sometimes the boundary fence, or wall, shows objectionably in an opening. In such case an earth bank, turfed to the summit, may be raised in front of it, sufficiently high to conceal it altogether (*as fig. E*).

The lawn is almost the most pervading landscape feature, and by its beauty it sweetly expresses repose. It is quiescent; it is never agitated by angry winds; its charm is modifying, and is an essential in the nature-picture. The turfed area about a house, and it may be said all turfed ground within the garden enclosure, is generally termed lawn; but the lawn is properly a grassy expanse in or between woods. The lawn of our garden should so far be true to the definition that it should present, whenever that is possible, the appearance of a natural opening, or grassy glade in a wood, where the sides are closed in by trees, and the distant end is lost in forest gloom. We may not at first have in our garden forest trees to shut in this glade; but we may arrange groups of planting to border our lawn, and the view may be directed across the park till a broken and undefined ending is reached. For such an effect, plainness and simplicity must characterise the treatment adopted. Formal beds for flowers or plants, to break the grand sweep of grass, must be avoided. A lawn should naturally be in the lower ground; specimen plants may be planted about the sides, preferably on the higher ground, so that the eye may be led to dwell on certain points, and to unconsciously pierce in particular directions; and particularly in order that the shadow of the trees or shrubs may fall athwart the grass in due season. But the sweep of the lawn should be unimpeded and clear, and nothing should invade its expression of repose. As it is not desirable to have too spacious an area of turfed ground, lawns will most generally take the form of vistas, and will serve chiefly to give a background of rest to the various features displayed.

Excepting formal slopes, all turfed ground, as well as all surface curves in general formation, must take us from one level to another, not by one inelegant drop, like the edge of an inverted saucer, but by a double curve commonly called Ogee (*fig. H*). It is without question desirable to have views so arranged that they may illustrate the extent of the place, and confer, if only by appearance, the sensation of spaciousness; but, in regard to the garden proper, it is most undesirable that its extent should be visible at a glance from any point. The spectator should receive or retain an idea of its vastness, by reason of the contrasting treatment of its several divisions, each insensibly separated from the other, but not screened off, while every point of view is carefully considered in relation to the desired effect; and this effect, as well as that of the house and its surroundings, must have its artistic and natural relation to the landscape. Incongruity and inappropriateness are inimical to harmonious, sweet influences. Things beautiful in themselves are destructive of beauty when inappropriately placed, and particularly of that beauty which should come to us in a sweet, indefinable effluence from a landscape near or far.

# WORK OF FORMATION.

---

THE foregoing observations on general formation have related to the plan. The following concern the practical work for carrying the purpose into effect.

Firstly, if the general plan showing a proposed arrangement of the house, its terraces, pleasure gardens, fences, groups of planting, drives and walks, its kitchen garden, glass-houses, etc., be designed for uneven ground, and particularly if an estimate of expenditure be required, detailed plans must be prepared, showing the intended levels. If a new place has to be made, it is advisable to strip the surface soil from the site of the house, and of the ground immediately around it. Some, if not all this soil should be run into a convenient heap for future use near the house when the structure is raised, and if there is any soil not needed for this position, it will be available for plantations. When the ground is falling, or uneven, it is advisable to form it roughly about the site before the builders commence their operations. Builders, when once they begin, are apt to spread their material over the ground, and frequently do not remove it till the conclusion of their work. All trees and plantations that remain close to the building must be protected. The next operation must be to make some portion of the approach, preferably only the back road to the house, which can serve the use of the workmen. Levels of all drives should next be taken on numbered pegs, at a minimum distance of 100 ft. apart, and the gradient line must be marked on the plotted section; that will give the depth of cutting or filling, at the position of each peg. It is profitable if you can so arrange the gradient that the soil from a cutting is sufficient for all filling required without length of lead; yet careful regard must be given to treating the gradient, so that earth may be obtained at any particular point for raising mounds for planting, or for other purposes. Some difference in the slopes and gradients for the approach drive, and for the back drive, should be made; the former should be flatter and more finished than the latter. When the levels have been settled the drain can be put in; the capacity of the drain must depend on its length, and on the quantity of water it will have to carry off. If only surface water from the road is to be conducted, 3 in pipes are sufficiently capacious for a distance of 200 yds.; but for a longer distance 4 in. pipes are needed. In estimating the quantity of water that is to be provided for, regard must be given to the surface drainage which may come on to the road from falling ground above, unless the general drainage be quite independent.

The drain is best placed in the centre of a road, with gully holes wherever necessary for the gradient. Drains may, however, be placed at the sides, and they should be filled up with porous material. Gully holes on drives should be built in mortar; though in garden walks they may be built "dry"—that is, without mortar. In long drives through parks, drains are frequently dispensed with, and the water turned on to the grass land at convenient intervals. On drives, 9 in. gratings and frames are generally sufficient; for garden paths, 7 in. gratings are large enough. On perfectly level roads, or on steep gradients, say from 1 in 10 to 1 in 15, gully holes should be 25 yards apart; but on a slight incline they may be as far apart as 50 yards. On walks they should be more frequent, because the gradients are steeper, and the gravel used being finer, is more easily displaced by rain; the walks also intersect, and are more undulatory. Outlet pipes from gully holes should be fixed at least 6 in. from the bottom, to allow of that space for possible deposit, which can be cleared out as occasion serves. In forming either drives or walks it is best to make the verges or sides first, as a good line is most readily obtained in this manner. The verge should be built up with turves laid on edge if the gradient shows filling. If turves are not available, let the ground be well trodden till it is solidified. A couple of inches should be allowed for subsequent cutting when the true line of the drive is neatly trimmed. Care must then be exercised, not only that the true line is cut for exact width of the drive, but that the cutting is carried to full depth, and the depression can be filled with stones. At least 6 in. of hard, porous material should be spread over all drives, and 4 in. on all walks. This covering may consist of broken bricks, or stones, or coarse gravel, or well-burnt ballast. On that a coating of stone that has been broken to pass through an inch ring, or of fine gravel, 3 in. thick on drives, or 2 in. thick on walks, should be laid. Drives, where two carriages may pass, must be at least 14 ft. wide; for one carriage, 9 ft. wide. Walks should never be less than 5 ft. wide; ordinarily, they should be 6 ft. wide. A drive of 14 ft. width should round over 3 in.; a walk of 6 ft., 2 in. The turf edging should rise 1 in. above the rolled surface of the road or walk.

The terraces should be formed thus: First strip the soil from the site. When the filling is over 3 ft. in depth it should be done in layers, and each layer consolidated. In any case, the ground should be well rammed as the earth is deposited, and when it is possible, the ground should be given time to settle before the turf is laid. All terrace slopes, as they have ordinarily a southern aspect, should have 9 in. of soil on them, to obviate the danger of the turf "scorching." All slopes and angles must be perfectly formed to receive the turf. The turves should not be rolled over the angles, but be laid to, and join, at them, so that a sharp arris be secured. On the site of any beds for flowers, or foliage plants, the soil must be of proper quality and sufficient depth, and when deposited in solid ground should be drained. Grass is cut into turves 3 ft. long and 1 ft. wide; they should be about 1 in. thick. When such turves are rolled up, they will in autumnal and early spring weather last for several weeks without permanent damage. Before cutting turves, the grass should be mown, and then well rolled. As they are relaid, all daisy, plantain, and bad grass roots should be taken out. Care is required in cutting turves to regulate the work so that the sections of grass may not be removed, or accumulated, save as the site is prepared to receive them; and the surface made for them should have 6 in. of loam. When the ground is to be sown instead of turfed, after the site is prepared, let the sowing be made in springtime, and during showery weather. A good mixture of seed for a lawn is that recommended by Mr. R. Thompson, as follows (the kinds of grass are those suitable for forming a good lawn, and the quantities of seed given are those required for an acre):

| | |
|---|---|
| Lolium perenne tenue | 20 lbs. |
| Cynosurus cristatus | 5 ,, |
| Festuca duriuscula | 3 ,, |
| Festuca ovina tenuifolia | 2 ,, |
| Poa nemoralis | 2 ,, |
| Poa nemoralis sempervirens | 2 ,, |
| Poa trivialis. | 2 ,, |
| Trisetum flavescens | 1 ,, |
| Trifolium repens | 6 ,, |
| Trifolium minus | 2 ,, |
| | 45 lbs. |

The above proportions will be found suitable for the generality of soils; but in shady situations to which *Poa nemoralis* and its varieties are particularly well adapted, 1 or 2 lbs. of each of these may be substituted for *Trisetum flavescens* and *Festuca tenuifolia*; whilst in rich heavy soils, 2 or 3 lbs. of *Cynosurus cristatus* and 1 lb. of *Festuca duriuscula* may be substituted for *Trisetum flavescens*, the quantity of *Trifolium minus* being reduced to 1 lb.

A single tennis lawn should have its level at least 100 ft. long by 50 wide. It must be well drained. If to obtain this level place excavation has been made on the upper ground to fill up the lower, a drain should be carried along the foot of the upper sloping ground, and be filled in with ballast, in addition to drains laid across the lawn. The soil laid on the surface of a tennis lawn or cricket ground should not be too deep or too rich. Six inches of moderately good soil is sufficient, and, except for light soils, a thin layer of burnt clay or ashes should be laid 6 in. beneath the surface stratum, and this layer of ballast should connect with the drains, so that quick passage of the water is ensured.

For all plantations the sites should be doubly trenched, that is, dug two spits deep, and the ground beneath broken up. The difference in growth of trees planted in trenched and untrenched ground is very conspicuous, and the value of the former is so much enhanced that in four years the cost of trenching is repaid. When the planting is to be on a raised mound, first remove the top spit of soil and place it aside; next trench the ground from which it has been taken, and make the mound; then use the soil reserved from the upper spit to form the surface and summit of the raised ground. There are many ways of obtaining earth for such work, and although some observations on the subject have been made in treating of the plan, it may be well to enumerate them here. The object will be to obtain earth for raising the ground, not only for producing effect, or promoting the growth of plants, but for hiding boundaries or masking unsightly objects that are irremovable. Ground that is taken near at hand is most convenient, and it is cheaper to work it by barrows than to bring it a distance in carts. Such earth may be obtained from the foundations of the building; by lowering the ground at foot of the terrace; by making the lawn at lower level; by lowering the surface between trees or places intended for single plants, or between larger groups; by removing ridges; by making hollow sunk panels, descending rose gardens; by lowering the lines of walks in certain places, so that these, in addition to providing such required earth, may add, by their undulating gradient, to the beauty of the landscape. Top soil should always be spared, and subsoil used for general filling, and this last so arranged that 2 ft. of top soil should be available for the planting area.

In lowering a lawn — say 1 ft. — the method of work should be what is termed back-handed; that is to say, the top soil of a narrow strip across a given area should be taken off and laid aside; then a foot deep of the underlying soil should be taken out and wheeled to the required place of deposit; then a similar strip should, in like manner, have the topmost foot of earth taken off and cast on the already lowered first strip, the new surface of which being first broken up; this operation being repeated with each succeeding strip of the space, till the last is reached, when the top soil laid aside from the first division can be deposited to make good the surface. The earth thus gained remains for use.

One inch of rainfall per acre gives 101 tons or 3,630 cubic ft. of water, which would fill a tank 30 ft. by 12 ft. by 10 ft., yielding 2½ cubic ft., or 15¾ gallons per minute, if the rainfall continue 24 hours. The annual rainfall in London averages 26 in., and it is reckoned that two-fifths of it percolates into the soil. It is by access of oxygen conveyed by water that plant growth is stimulated, and rain-drops are highly oxygenated. Plants will not grow in ground saturated with water from which the oxygen has been more or less taken up, and drainage is therefore necessary in promotion of vigorous vegetable growth. Such drainage may be consequent on natural formation of the ground, or be secured by the removal of obstacles, or by artificial means. But the cold stagnant water that has, as it were, served its purpose must be removed, or conducted to where surrounding conditions admit of its return by evaporation to discharge its function again in the rainfall. A circulation of warm air follows the descending rain which percolates charged with many ingredients for the sustentation of plant life. Water is heaviest at a temperature of 40° ; heat cannot pass downwards through water; hence the saturated ground can only be renovated by sun heat on its surface, and the dense cold water should find its way from the subsoil by drainage, either natural or artificial. In Denmark they practise a plan of double drainage; a system of piping serves ordinary purposes near the surface, but these drains communicate with a lower system, which again is generally conducted to some natural water storage, the outlets into which are guarded by sluices. In seasons of drought these sluices may be closed and the water retained. The soil there is generally light. The depth at which drain-pipes should be laid, and the distances at which they recur, must be ruled by the nature of the soil. In stiff clay they should be placed 3 ft. below the surface, and 16 ft. apart; in light soil they may be put at intervals of 40 ft., but 4 ft. deep. In non-porous soil, cuttings for the drains should be filled to within 6 in. of the surface with porous material. All main outfall drains should be laid with a minimum fall of 1 in 500, or 10 ft. 6 in. per mile. Collecting drains should have a minimum fall of 1 in 250, or 21 ft. per mile. These should enter the main ducts at an acute angle. In any system of drainage it is desirable to make wells at convenient places, that may serve as catch-pits, and inspection chambers. The direction of main drains should be as closely as practicable that of the lowland, and side drains should follow the slope in parallel lines. The outfall of the main should be above the water level. Ordinarily for a garden the injunctions just made do not so strictly apply. They are modified by important considerations. The walks are generally used for lines of main drainage; these in most cases traverse ground of steeper inclination than the open land, and there is an obvious variation in the minimum ratio of fall as set out above. It is not advisable to take any main drain nearer to the trunk of an established tree than the limit that is covered by its foliage, because it is not well to alter its former water supply.

## PLANTING.

---

THE word planting, which means the act or process of planting, has been adapted with an extended signification in connection with landscape gardening; it has been taken to indicate as well the thing planted, and is so applied in this treatise wherever the context determines its appropriateness. Planting, in act and in result, is a feature of paramount importance in the art. The well-directed practice or the misuse of it is essential to, or fatal to, the development of those aspects and influences of beauty, which it is the artist's purpose to conserve, to emphasize, or to induce by natural operation. It is a principal means in his hand. As a skeleton is clothed in beauty of line and colour by the flesh, so the undulations and sweet or grand features of the ground in its formation, are clothed by the all-pervading colour of the grass, with its delicious expression of restful quietude; and thereon Nature lavishes her objects of infinite variety in form, height, colour, in gradations of tone, and splendour of contrast—the absolute infinity of curve in detail, which exercises a subtle charm on us, that would deteriorate could we define it; the gradations of the sky-line, the fine irregularities of line where the lower verge of foliage touches lawn or grass land, the emphasis of sunshine and of shadow—and her gifts in this wise are trees and plants. These, then, each beautiful in itself, become chief means by which the designer works in forming the landscape.

A hill is made to appear higher if its summit be planted. A grand irregularity is given to the sky-line. When the hillside is clothed with foliage, then the variety of form, of colour, of light and shadow, are brought into play, and should be profoundly studied for the production of those subtle effects of beauty, the fascination of which we feel without the care of definition. The infinity of form and of tint gives variety of effect. If the foliage were so disposed that formality intruded here, regularity in the height of the trees, similarity of form, and an appearance of massiveness in the leafy clothing of the hillside, how wasteful of opportunity for beauty of effect the planting would appear! When such a regularity in foliage is found, it should be our object to break up its apparent surface. We should never create it. The effect may be narrowed to a single tree. Notice how inferior is the glow of light on an unbroken mass of leaves, when compared with the same rays falling on a tree whose boughs are irregularly branched, as is an old oak; the illumination touches the tips of the branches, qualifying their tints, which tints appear varied in half shadow, and that is set off by the almost gloom of the deep recesses; indeed, there is an expression of mystery in the picturesqueness that is presented which augments the beauty of the object. It is the same when the single tree is part of a grand assemblage of them, or when he towers above his compeers. The artist can use the variety of form in various trees, with the gradations and contrasts of colour they give him in varying conditions, almost as a painter uses his pigments. When the hill is grandly irregular in its natural form, then the effects to be aimed at, in present realisation or in future years, are proportionally more easy of attainment. The landscape gardener must always have his vision projected through the coming years that will develop his scenery by growth; and he must realise in imagination, as he contemplates a rugged prospect that he has newly or partially planted, how it will one day present the

> . . . Magnificence of many-folded hills,
> And promontories set of solemn pines!

This indispensable looking forward need not impair our present means of creating beauty. There are shrubs and trees of large size to be procured that can be carefully transplanted without danger of loss. When trees of small growth are of necessity used, temporary plants, acting as nurses, serve not only to promote the growth of the permanent plants, but also to cover the ground and relieve the monotony of low planting. Groups of trees on artificial or natural eminences should be most carefully selected in regard to the height of their ultimate growth, as well as their outline form at maturity, their juxtaposition, and the contrasts so to be created in enhancement of their beauty, not only as individual specimens, but as component features of the group in which every part has relation to the whole. One paramount use of planting is to hide what may be unsightly or incongruous in a prospect; or, by a well-arranged break in the procession of trees, to admit some beautiful view, or some appropriate object, such as a church spire, or a glimpse of water, to the vision of the spectator at a particular point where the picture presented may be most grateful. Such a picture may be, as it were, framed in the grand forms of the foreground trees. A most important use of trees in the general design is to connect the different parts of it, by continuity of planting and by repetition of tint and form.

In planting a great place there should be a distinctly apparent difference between the garden, or the enclosed space round the house, and the ground beyond that stretches into the park. The distinction can be maintained thus: Near the house the planting should be finer, and of trees and shrubs that are not indigenous; or, rather, such trees and shrubs may be abundantly used there, to produce their effect; for the introduction into a district of plants not indigenous to it marks an innovation, as it were, and shows the hand of man. Such plants or trees must be sparingly used, or not at all, in the park, where an object is to conceal the fact that all is not due to nature in its local development. Both the character of the trees and shrubs, as well as the disposing of them, should not militate against this idea. The plants used in the garden should be those that in their growth, tint of foliage, and colour of flower, are delicate, rare, of neat outline, and such as we are accustomed to connect in idea with the garden. The more varied sizes of the groups will conduce to a like impression. Such association of idea with particular plants is necessarily a fanciful and changing one, because of the frequent introduction of hardy trees and shrubs from other countries to be naturalised here; but it nevertheless plays an important part in landscape gardening.

In the garden we should group the plants with regard to sorts, to colour, to form, and to size; but such groups will necessarily be small, and they should admit of a view of the individual plant. This particularising of the individual plant differentiates the planting of a garden from that of the park and plantations beyond. In the outside plantations, massing takes the place of grouping, as the effect to be produced is generally realised from a distance, and gives con rast in colour and shape. In the garden we have, to a certain extent, a collection of plants; in the park we have plantations. In the former, a grouping of plants should be made in different positions; and though bare repetition, or balanced uniformity, should be sedulously avoided, yet the several clumps should have an inoffensive relation one to the other that should illustrate the oneness of the composition.

The old saying, "Plant the hills and bare the vales," is, like most proverbs, right in the restricted sense of an obvious application. It holds generally good as coinciding with the natural arrangement of the formation of the land surface by action of water; but a place would appear very dotted and formal if that mode of treatment only were adopted. If the primal natural operations gave this formation of the ground, and vegetation was first on the hills, yet, as the soil washed from the hillsides to the valleys where the rivers flowed, vegetation soon crept along either side of their course, and was nourished there. If we copy Nature's operations in this way, we find her planting on every side—not only on the hills. We should seize on the spirit of the beauty in her operations, and, with such license as her developments allow, foster the expression of her delightful influences, and train them to our use. In the restricted scenes touched by the landscape gardener, we have, by the introduction of foreign plants, the power to create, on a small scale, the beauties of plant life in all temperate climes; our selection depending on soil, subsoil, elevation, distance from the sea, and latitude.

There are three points of importance to be determined in regard to groups of planting, viz., the site, the ground outline, and the disposing of the trees and shrubs. The conditions under which the design has to be created, and applied, vary in almost every case, and no hard and fast rules can be made for adoption. In one place the artist's design may be modified by the necessity of creating shelter, in another the planting may be needed to mask unsightly objects that are near, in a third to direct the gaze to distant beauties of landscape that may be brought into the whole effect, like possessions. Each place has its peculiar considerations. All that can be done here, in this connection, is to speak in general terms, and to particularise only with certain groupings that may illustrate the general question.

When the ground is broken or undulating, the advantage should be seized of marking eminences by planting. Rising ground may be in appearance raised still higher if we cover it with wood. The trees should be tallest as they reach the summit. Trees standing singly mark and emphasize falling ground. Plantations crossing falling ground generally are inharmonious to the broad effect. The brow of an eminence should not be seen above trees; and if the brow forms a tedious continued line, it should be broken by clumps or large masses along its range, and by dividing the line into very unequal parts. Openings thus created may be treated by placing at the sides of them, at all events not in the line of vision, various small clumps, so as to train the spectator's glance in the given direction, and to promote the idea of distance. The outlines of groups in a park should not be regular, or in a succession of easy sweeps, or form a serpentine line. They should have strong prominences marked by detached trees that stand boldly in the group, and they should present deep recesses. Exactness in outline will not be preserved; but if the general features be well set out, natural alterations due to the growth of the plants will not spoil the effect. In plantations near the eye, lights and shadows are more apparent than on distant groups of trees; the effects are stronger; therefore dark foliage planted in a near recess makes it appear deeper still. White foliage, or blossoms, seem to be nearer to the spectator than they really are. Objects become fainter in effect as they retire from the eye. A detached clump, or a single tree of lighter green, will therefore seem to be further off than an equidistant planting of darker hue. A regulated gradation of one tint to another will apparently alter the length of a continuous plantation, according as the gradation commences with foliage of light or dark green. When a long continuous line of planting fronting the eye cannot well be broken, because perhaps it may serve as shelter, or for some other reason, a variation of line and composition may be given to it by placing light green plants in graduated tints at certain positions in its course. The effect is to give the idea of distance at such points, and this effect of expanse will be still more promoted, if the sky-line of the tree-tops is lowered with a gentle curve lowest where is the lightest shade of green. Among the dark-foliaged trees may be particularised the Oak, Chestnut, Beech, Elm, Austrian Pine, Spruce Fir, Cedar of Lebanon, Hemlock Fir, Yew, Holly; among the light green, Plane, Birch, Ash, Acacia, Lime, Poplar, Willow, Laburnum, Larch, *Abies Concolor*, *Pinus Insignis*, *Abies Douglasii*, *Cedrus deodara*; among the red greens are varieties of Maple, the American Oak, Copper Beech, etc.; among white greens are Poplars, Sea Buckthorns, etc.

An avenue is the assumptive expression of grandeur—of grandeur that must be achieved at its starting-point, belong to its course, and be conspicuous at its ending—an essential feature of its purpose. It must never be petty or inconsequential in these respects. It should be planted only when importance is to be given to its line, and when an imposing ending can be bestowed on it. Its line should be, if possible, straight, and free from extraneous planting, unless, indeed, it pass through a wood, and form part of it. Its trees should be considered as belonging to, and forming part of, the road. It is essentially a stately feature in its appearance; and its purpose should be harmonious with its expression. The mind is apt to resent a paltry result to such promise as it impressively gives, if the ending be unworthy of the beginning. When age and luxuriant growth come upon it, there is a venerableness in its beauty that is not otherwise obtainable by any like means. It has a charm and dignity of purpose when young, if well planned, that is soon apparent. But when ill planned—to conduct the vision or the footsteps from some point of little importance to a position of less interest—it bears all the penalty of a wasted resource; and the misapplication of such means reacts on the imagination, and the feeling of the spectator, with adjusted force. It must be remembered that an avenue across a park divides it in two parts at that point. It is not fitting except it proceed from some important entrance on a side whence no distant views can be spoiled by its introduction. The width of a grand avenue should be 50 ft. The trees may be, preferably, Elm, Beech, Oak, Chestnut, and they should not be planted nearer in precession than 40 ft., unless they be planted at intervals of half that distance with the purpose of destroying alternate trees, as their growth makes the removal necessary.

A century ago it was the fashion and practice to make boundary plantations continuous and unbroken on the verge of an estate. That is a practice not to be recommended. It is well to place large masses of foliage towards the boundary, especially if these are arranged to present winding intervals of turf, and so that endings and corners be masked; or such planting may be made only for effect. But an enclosure pure and simple, even though it be of leaves, and not a brick wall, gives a shut-in and cramped feeling, which needlessly militates against expressions of beauty and expanse that may be deftly gained from outside the boundary line.

Groups of planting in the garden may be more apparently connected with the design than is permissible in the park. The effects are for nearer contemplation, and are more intricate; they harmonise with, but display, the general design. The seeking after natural formation is restricted, the reproduction of it more in miniature, the arrangement is more elaborate and precise. Our choicer plants have to be displayed here to best advantage, and the mean has to be obtained between a mere collection of rare trees and shrubs, and such a grouping of them as will produce a harmonious polytonous effect as a whole. To achieve this, the outline of different groups, and the position of the plants in them, give us requisite help. By the outline of a group is not meant a hard artificial boundary, enclosing an earthen bank on which trees and shrubs are planted; but a line to which it is intended that the plants should spread, covering the earth surface, and such an outline should be graceful, and give the required filling to the general design, even when the verge first marked may be overgrown. The outline of such groups in a garden is to be set out, and to it the ground is generally turfed. Nothing can be in worse taste, or more in derogation of the purpose of such groups, than to mark the outline with a formal row of bedding plants, or other stiff edging—a mistake that is frequently committed. Where low flowering plants are introduced, they should grow naturally to the edge, and separate plants may be placed beyond the main group, so as still more to destroy simple formality of effect in this connection. Formal planting may be introduced with propriety at the intersection of a walk, or at a point where there is some artificial combination; but unless quite appropriate, its best effect is dearly purchased by the loss of naturalness in expression that may otherwise be achieved or maintained. When formal bedding-out is made thus at particular positions, the containing bed should receive a regular shape, and be made with flat surface, to distinguish it from the natural grouping. In other regions of the place, formal arrangement and bedding-out is quite appropriate—on the terrace gardens and parterres, where plants are brought from the glass-houses for seasonable growth and decoration, or where permanent evergreen foliage plants are arranged, for instance. There is, moreover, the planting of the so-called "American" garden to be considered, where plants delighting in a peaty soil are grouped together, such as Rhododendrons, Azaleas, Kalmias, Andromedas, etc. There is the trim yew-hedge, lined with recesses for vases and for seats, in close proximity to, or surrounding the old-fashioned herbaceous garden. Another division is the rock garden, where the blossoms of Alpine plants glow in their varied colours. The Pinetum, where are collections of hardy conifers, each tree rising separately from the lawn, with a distinctive beauty, which not only charms the eye but creates a special interest. Then we may, perhaps, consider the low ground toward the lake, where the planting may be characterised by introducing a collection of damp-loving plants, beyond which, on the side of the lake itself, will come the collection of weeping trees, overhanging the water surface. These are all special positions, and each has a special planting, and its relation to the whole decorative effect. The disposition of the plants requires skill. In addition to the grouping for foliage, colour, and height, which has been insisted on above, there is another consideration. It is pleasant to find in a group a collection of various kinds of trees of one species—for example, of Hollies, having to lighten their effect an undergrowth of variegated Dogwood; the vacant spaces in the first instance can be temporarily filled with common Holly. Or a collection of *Retinosporas* may be made, interspersed with *Spiræa callosa*, etc.

Assuming that the ground for the groups has been raised, as already recommended, the arrangement of the planting should follow the indicated level of the bed; the points should receive prominent trees, and the lower spaces should be filled with less important plants, more dwarf in habit. Those trees and shrubs which are hardy or delicate should be allocated to positions where they may protect, or be protected. By grouping is not meant a repetition of planting by threes or fours, but that some trees of the same species differing in height be planted in juxtaposition to trees of other species having differing form and colour that may harmonise and at the same time contrast. Equally, variety in colour is not to be taken as meaning the repetition of violent contrasts, such as Copper Beech and Variegated Maple, so frequently planted together, with the erroneous idea that strong contrast is all that is required. In mass-

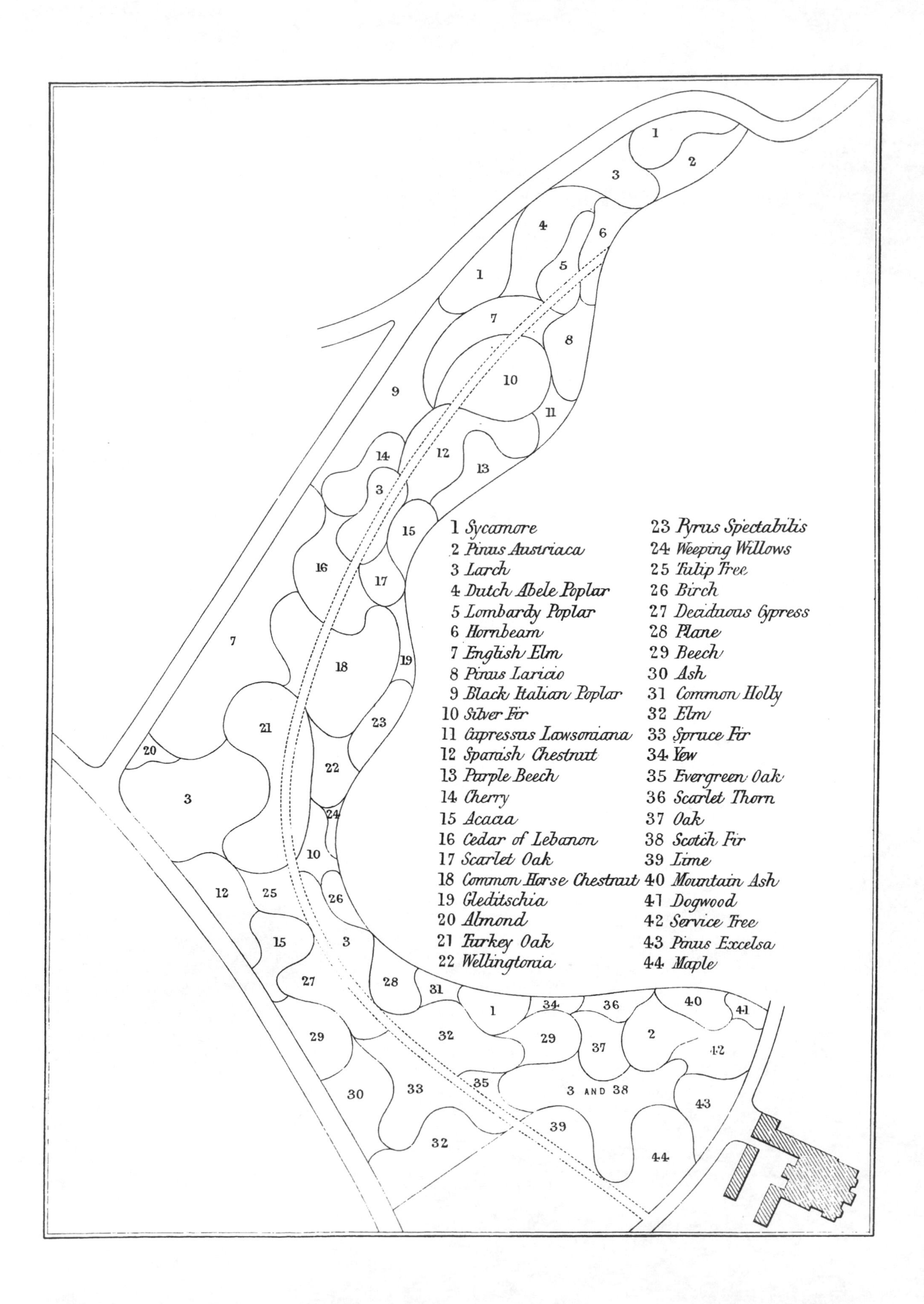

1 Sycamore
2 Pinus Austriaca
3 Larch
4 Dutch Abele Poplar
5 Lombardy Poplar
6 Hornbeam
7 English Elm
8 Pinus Laricio
9 Black Italian Poplar
10 Silver Fir
11 Cupressus Lawsoniana
12 Spanish Chestnut
13 Purple Beech
14 Cherry
15 Acacia
16 Cedar of Lebanon
17 Scarlet Oak
18 Common Horse Chestnut
19 Gleditschia
20 Almond
21 Turkey Oak
22 Wellingtonia
23 Pyrus Spectabilis
24 Weeping Willows
25 Tulip Tree
26 Birch
27 Deciduous Cypress
28 Plane
29 Beech
30 Ash
31 Common Holly
32 Elm
33 Spruce Fir
34 Yew
35 Evergreen Oak
36 Scarlet Thorn
37 Oak
38 Scotch Fir
39 Lime
40 Mountain Ash
41 Dogwood
42 Service Tree
43 Pinus Excelsa
44 Maple
3 AND 38

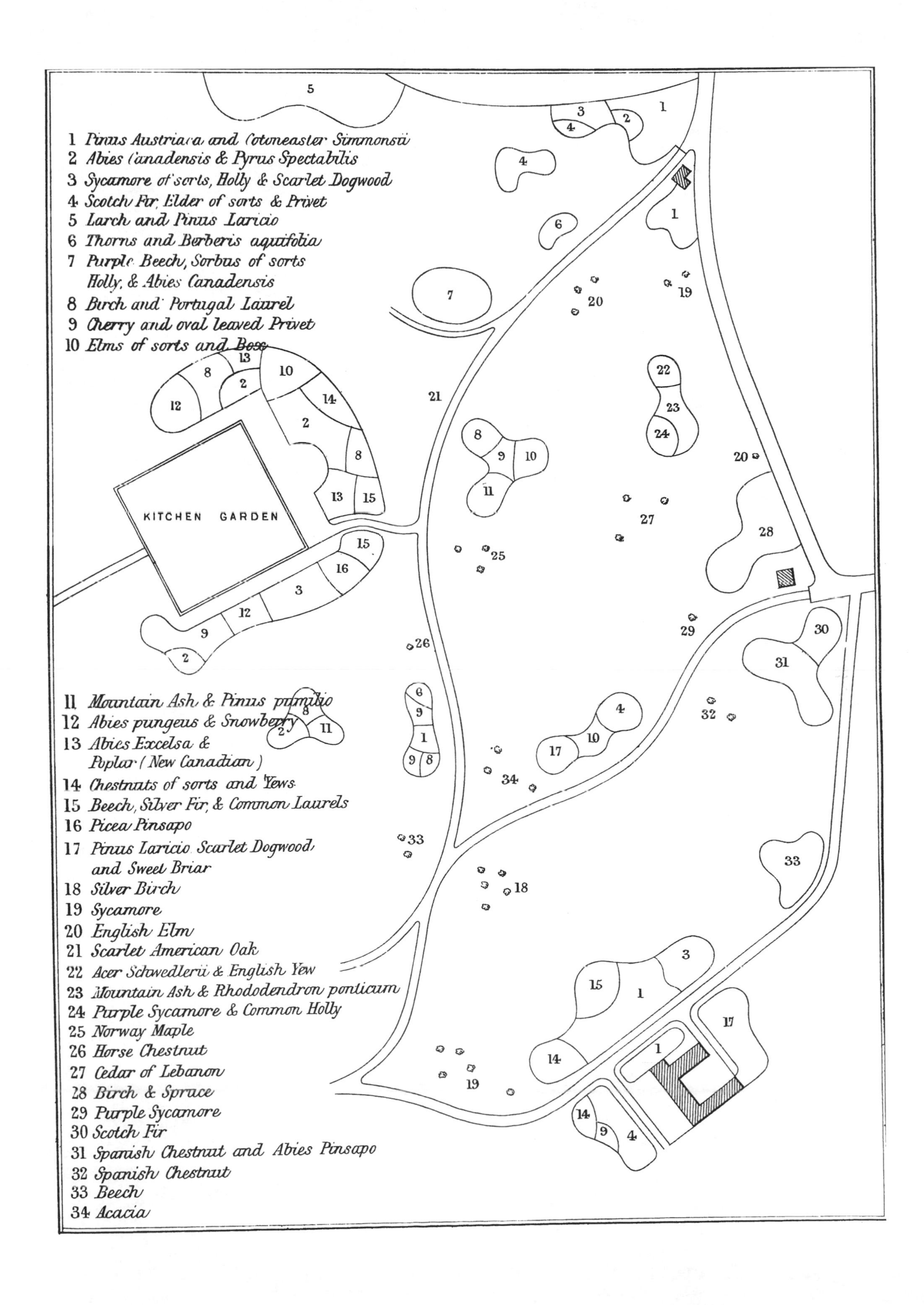
1 Pinus Austriaca and Cotoneaster Simmonsii
2 Abies Canadensis & Pyrus Spectabilis
3 Sycamore of sorts, Holly & Scarlet Dogwood
4 Scotch Fir, Elder of sorts & Privet
5 Larch and Pinus Laricio
6 Thorns and Berberis aquifolia
7 Purple Beech, Sorbus of sorts
Holly, & Abies Canadensis
8 Birch and Portugal Laurel
9 Cherry and oval leaved Privet
10 Elms of sorts and Rose
11 Mountain Ash & Pinus pumilio
12 Abies pungeus & Snowberry
13 Abies Excelsa &
Poplar (New Canadian)
14 Chestnuts of sorts and Yews
15 Beech, Silver Fir, & Common Laurels
16 Picea Pinsapo
17 Pinus Laricio Scarlet Dogwood
and Sweet Briar
18 Silver Birch
19 Sycamore
20 English Elm
21 Scarlet American Oak
22 Acer Schwedlerii & English Yew
23 Mountain Ash & Rhododendron ponticum
24 Purple Sycamore & Common Holly
25 Norway Maple
26 Horse Chestnut
27 Cedar of Lebanon
28 Birch & Spruce
29 Purple Sycamore
30 Scotch Fir
31 Spanish Chestnut and Abies Pinsapo
32 Spanish Chestnut
33 Beech
34 Acacia
KITCHEN GARDEN

ing tints attention must be constant to realise the form of the tree as well as the colour of its foliage, or the trees may carelessly be so arranged as to present the appearance of strips one beyond another. The group should rather be divided into considerable portions, each of a rounded figure, to be filled in with plants that will give the various tints required, and are of the necessary height. Contrasting tints, such as dark and light greens, coming together in large masses of planting, break the surface, so to speak, on which they meet. An outline which cannot for any reason be varied sufficiently in form, may be diversified in appearance by clever management of such shades. Every opposition of shade will vary a continuous line. When a distinct hollow is apparent between two clumps of trees, it is sometimes effective to plant the sides of each group next the hollow with trees of similar kind, so as to create or suggest an idea that the groups have been separated by the cause that made the hollow.

In addition to the clumps of planting made for effect in a park, there are to be considered those that are to have as well a further use, namely, to afford cover for game. For the flight of a pheasant these should be placed about 200 yds. apart. Among the trees should be planted thickly Cotoneaster Simonsii, Rhododendron ponticum, Scarlet Dogwood, Gorse, Broom, Black Thorn, Snowberry, *Berberis aquifolia*, Sweet Briar, Box, Privet, Elder, etc. An example is given showing the arrangement of trees and shrubs in outlying park plantations (*see Plan*).

In planting a villa garden there is a different treatment to be observed from the stocking of a large place. In the small garden it is generally necessary to secure an immediate effect, with an expression of completeness that shall appear in winter as well as summer. Shrubs that are of dwarf habit or of slow growth, if purchased of sufficient size to break the monotony of low planting, are generally expensive, and the choice of sorts is somewhat restricted. Recourse is therefore frequently had to plants, chiefly evergreen, which, though looking well for a time, grow in no short period beyond the needs of a small garden, and by their shade injure the growth of rarer shrubs. It is, however, mostly necessary to use some fast-growing trees; but the best plan is so to arrange the general planting scheme, that such trees may be cut down and removed after serving their purpose. There are, however, many plants that attain a great size with sufficient rapidity, such as *Cedrus deodara* and *Atlantica, Cupressus Lawsoniana, Thujopsis borealis* and *dolabrata, Thuja* of sorts, Portugal and common Laurel, Hollies, Limes, Acacia, Beech, etc., many of which, by constant pruning, may be kept within size fitting for a small garden.

The moving, or transplanting, of large growing trees is a resource of great utility, and can be carried out with less difficulty than is frequently incurred by processes ordinarily employed. Some of the methods resorted to, in general circumstances, create much of the difficulty they are designed to overcome; necessitating the use of expensive and elaborate machinery, when a much simpler process will serve. In practice, I adopt the method described later on; I have found it easy, inexpensive, expeditious, and successful. With a staff of 30 men, as many as 20 trees, whose stems average 1 ft. in diameter at the ground level, with a height of 30 to 40 ft., have been thus safely removed to new positions, at an average distance of 100 yards, in one day. The trees were prepared on the previous day. Trees can be moved in this way over hill and dale, and on to newly-made mounds, where it would be difficult to employ a tree-moving machine dragged by horses. In dealing with an estate it is frequently found that trees have been long planted in belts at the boundaries; or they are in the way; or are wrongly placed; or, perhaps, are crowded together, to their detriment. It is necessary to remove some, and they can be shifted for use elsewhere, instead of destroying them. They form most valuable features in a newly-made place, securing immediately effects that would otherwise have to be long waited for, besides conducing to visible beauty. The most suitable trees for such removal are Horse Chestnuts, Sycamores, Maples, Limes, Beeches, Service trees, and Mountain Ash.

The method to be adopted is as follows: Open out a trench 5 to 7 ft. distant from the stem of the tree, such distance depending on the size of the tree, and dig down, cutting off any of the larger roots that may be met with, but carefully preserving all the smaller fibrous roots, following them to their ends and laying them carefully round the ball of the tree. At a depth of from 3 to 4 ft. begin to dig nearer the stem, and shape the ball by removing any superfluous, loose earth at the top and sides, so that the root-ball has somewhat the appearance of a flat inverted cone. Leave a supporting space of 2 ft. diameter under the stem, then place matting round the ball, enclosing the fibrous roots, and rope this round firmly. Next, place thick planks on each side of the central support. Fix two or three guide-ropes two-thirds of the way up the stem of the tree. Pull one of these in the direction of the main planks, and draw the tree over somewhat, separating it from its supporting base, and at the same time, push two or three short 6 ft. planks as far under as possible; then pull the tree in the opposite direction and repeat the operation on the other side. The cross-planks are to be lightly bound together. The tree is then standing on a platform of planking, which rests, in its turn, on the two main planks first put under. These latter are the sliding-planks, or rails, on which the short cross-planks, forming a carriage, slide. Care will be taken that the sliding-planks are placed in the direction to be traversed, or in the direction of least difficulty in extricating the tree from its site. A piece of matting is then placed above the crown of the root, to prevent the bark being frayed, and a man, facing the direction in which the tree has to be moved, and behind the tree, takes a strong hempen rope in its middle, passes it over the protected stem, and drawing each end towards him, crosses the ball diagonally, and carries each rope, with a half-hitch, over the stage of cross-planks, and delivers an end of the rope to men on either side, who are awaiting it on the further side of the tree. These men, 10 to 20 in number, take the two ends of the rope and are ready to start. Planks are, meanwhile, laid from those under the tree to the required distance, care being taken that each succeeding plank-end is placed, for a distance of six inches, under the preceding one. These planks are then wetted and, if the way is uphill, smeared with clay. Two men hold guide-ropes to steady the tree, and the men on each rope drag it to the required spot without difficulty. If there are several trees to be removed, having a stem-diameter of 6 in., with a ball weighing about a ton, it is easier to move them on a made sledge, instead of the cross-planks; but the general operation is the same. The ground under the new planting site must always be broken up, and on the arrival of the tree, after the matting has been removed, the fibrous roots should be spread out, the tree regulated as to its best side and to its perpendicularity; and it should then be well packed with rammed soil, and the ground well filled in to the new surface. If many roots have been removed, the tree must be reduced proportionately. As a rule, it has been found wise to reduce the leaf area from one-fourth to one third. Branches should be cut where there is another ready to serve as a leader in the place of the reduced limb.

It would only be misleading to compile a list of plants suitable for stocking any and every garden, as each district varies in respect of its soil, subsoil, climatic influences, nearness to the sea, nearness to a town, its sheltered or exposed situation, and its latitude. Such conditions must be duly considered in working out the general planting arrangement of each place, independently of the grouping for effect. Nevertheless, it may be found useful to have a list of trees and shrubs suitable for special districts, and the plants, of which lists are given, are those that will grow, if not thrive, in

(*a*).—Shade, or can be used for undergrowth.

(*b*).—Chalky soil.

(*c*).—Peaty soil.

(*d*).—Exposed situations, or near the sea.

(*e*).—Smoky districts, or those in the neighbourhood of towns.

(*a*).—*Aucuba Japonica*, *Berberis aquifolia*, Bramble, Broom, Box, Butcher's Broom, Cotoneaster Simonsii, English Yew, Evergreen Privet, Elder, Holly, Portugal Laurel, *Rhododendron ponticum*, *Phillyrea*, Savin, Scarlet Dogwood, Snowberry, St. John's Wort, *Skimmia oblata*, *Thuja Wareana*, *Vinca*, *Abies pectinata*, *Abies alba*, *Abies Canadensis*, *Abies nigra*, *Pinus pumilio*, *Cephalotaxus Fortunei*, Black Thorn, Hornbeam, etc.

(*b*).—Privets in variety, Elder, common, black, and golden, *Veronica Traversii*, Currants in variety, *Leycesteria formosa*, *Buddlea globosa*, *Berberis aquifolia*, Box in variety, Laurel, both *Caucasica* and *rotundifolia*, Dogwood in variety, *Coronilla emerus*, *Cotoneaster Simonsii*, Thorns in variety, *Pinus Austriaca*, *Pinus nobilis*, *Daphne pontica* and *mezereum*, Beech in variety, Yew in variety, *Hibiscus Syriacus*, St. John's Wort, Hollies in variety, Poplars, *Prunus Pissardi*, Service tree, *Symphoricarpus racemosus*, Larch, Guelder rose, Periwinkle, etc.

(*c*).—Rhododendron, viz. named varieties, seedling varieties of *R. Catawbiense*, dwarf varieties, *e.g.* *R. Wilsoni*, *hirsutum*, *ferrugineum*, *myrtifolium*, *ovatum*, and *R. ponticum*, Kalmia, Andromeda, Azaleas, both Ghent and *pontica*, Heaths, Gaultheria, Ledum, *Menziesia*, *Epigœa*, American Cranberry, *Pernettya*, *Polygala*, *Rhodora Canadensis*, *Vaccinium*, White Cedar, Willow, Birch, Hemlock Spruce.

(*d*).—*Pinus Austriaca, P. maritima, P. cembra, P. insignis, P. pinaster, P. pinaster minor, P. pumilio, P. pumilio uncinatus, Abies Cephalonica, Abies Hookeriana*, Cornish Elm, Wych Elm, Poplars, Sycamore, Evergreen Oak, *Thujopsis borealis, Thujopsis dolabrata, Cupressus macrocarpa, Aucuba, Escallonia macrantha*, Box, English Yew, Holly, Double Furze, Sea Buckthorn, *Juniperus Virginiana*, Euonymus, *Phillyrea*, White and Spanish Broom, Groundsel tree, Bay tree, Berberis, *Rhamnus alaternus*, Privet, Laurustinus, Buddlea, Spiræa, Currants, Scorpion Senna.

(*e*).—Ash, Austrian Pine, Birch, Almond, Service tree, Horse Chestnut, Plane, Sycamore, Siberian Crab, Laburnum, Tulip tree, Thorns, Poplar, Willow, Aucuba, Andromeda, Azalea, *Berberis aquifolia* and *Darwinii*, Box, Broom, *Cotoneaster Simonsii*, Deutzia, Dogwood, Elder, Euonymus, Guelder rose, Hibiscus Syriacus, Hollies, green and Hodgins', Kalmia, Lilac, Laurustinus, *Leycesteria formosa, Philadelphus*, Privet, evergreen, oval-leaved, and Japanese, Pernettya, Rhododendron, *Ribes*, Sumach, St. John's Wort, Vinca, Weigela, Wayfaring tree, Ivies, etc.

# WATER.

Of all beautiful features in a landscape, none is more attractive than water; whether seen from afar, or contemplated near; in the form of a running, dimpled stream, or of a broad expanse reflecting the sky and nearer objects on its placid surface; when it spreads under the burning sunlight, when it glitters in the cool moonlight, or when it pulses and beats the shore, under a wind that bends the plants and trees on its margin all one way. Its beauty has a fascination of its own. A river has almost a personality in its district. When our steps wander, they soon tend to the river or the lake, and on the margin we seem to come near a friend. Wherever it is possible, and is fitting to do so, this means of beauty must be conserved and used by the landscape gardener.

A stream, or other water feature, has relation to the surrounding land, and its expressive character. A brawling brook, a rushing stream, leaping over a rocky bed, and foaming over rocky barriers, or sweeping between them, does not naturally occur in champaign country, and if artificially created there, is ridiculously unfit, as incongruous as would be any misplaced artistic feature. Water adapts itself to its superficial environment, and expresses its beauty with infinity of conditions. In utilising them, we should imitate the spirit of beauty in Nature's operations, but not distort her effects. The placid lake that may be formed on low-lying land, has a beauty and fitness there that are equal, in their way, to the charm of a mountain torrent; and the fitness to surrounding conditions is the measure of beauty for both. It would be as utterly wrong to imitate in a low, flat country the rushing stream that dashes in cascades over a rocky bottom, as in a mountainous district to work in simulation of a broad, sleepily-flowing river, which in a level region, brimming to the grassy verge of the fat meadows through which it slowly sweeps, is an object of particular charm. Lakes may, it is true, sometimes be properly formed in high ground, provided the inclination of the land is wrought to apparently lend itself to such a formation; and the question of fitness must be well considered.

The landscape gardener's treatment of a lake or a river differs. For the former space is necessary, and much of the pleasure derived from the object is due to its outline. This outline is gained by the arrangement of bays, creeks, and promontories. No figure perfectly regular should be used, not even a semicircular bay, which, though beautiful in shape, is not natural. The prevailing line must be concave. It is indispensable that portions of the lake and its shore be hidden from the spectator at other points on the margin. If the lake be large, the end may be turned behind a wood or a hill, or, failing such natural objects, by interposition of planting. A vast sheet of water may compel our wonderment, but the picture none the less is circumscribed, if we are able to take cognizance of its features too completely; our imagination is confused or frustrated. In such circumstances the mind seizes with avidity and delight on distant but definite objects, such even as a shore of broken form, or a headland. An artificial lake may be out of proportion to the size of the estate. If the shore be flat, the scene may be exceedingly uninteresting. In such case, by raising opposite banks, by planting trees, or by making constructions, an effect of proximity may be gained; for elevation, producing distinct-

M.

ness, gives that result; a low shore, without trees or other like features, though it may apparently increase the extent, does so by the diminution of picturesqueness. In practice, an apparent curved or bay-like outline may be given to a shore when viewed across a large sheet of water, by judicious arrangement of planting, and by irregular height of the banks. The extent of reflecting surface should be in proportion to the extent of the visible surrounding ground. If there seem to be too great a surface of water, and the outlying ground be flat, portions of the distant bank may be raised and planted; then the effect will be to give proximity to such objects, and to proportionally lessen in appearance the abundant expanse of water. Large trees on the opposite shore of a lake diminish in degree the apparent extent of the water. It is unwise to allow any shore, viewed across an expanse of water, large or contracted, to be fringed with trees, because they create a gloomy impression. If for no other consideration than that of admitting light, with its effect of freshness, an unbroken fringe of trees should not be maintained. In such case, the removal of some in particular situations, with a coincident lowering of the bank, will give an effect of lengthening the water area. Promontories or projecting points should be raised above the surrounding level. In forming a bay, not only should the ground be taken out where the water-line is formed, but the rising ground inland should be hollowed, and the surface be given a concave form, gradually diminishing in extent as it touches the water edge. Views from the house, or from settled points, should be directed toward the bays, not only because they give the largest prospect of water surface, or because of the lower pitch of the intervening ground, but because a bay suggests, with more or less appositeness, the idea that here was the line of torrent at the supposed natural formation of the lake, where the land was by such means denuded of trees. Lakes should be made to appear as the remainder and result of such a rush of water; the shores and surroundings should all bear the expression of having been moulded and modified by water action.

Islands serve to give variety to the outline of a lake; they can be used to mask the ends, and when they are made by high ground being left, a vast amount of excavation is saved. An island should not be placed in the centre of a bay, but to one side of it; and there should never be fewer than sixteen feet of water between any point of an island and the mainland. If a bridge be constructed from the mainland to an island, it should be at the narrowest point, and at right angles to the shore. When the surrounding land surface is undulated, water may enter a lake at its highest point by a waterfall or cascade, and the outfall may also be formed to rush over a ledge or brawl over a rocky bottom. There is an expression of propriety when water enters a lake by river-like means, and when it comes by a slight cascade, or a series of rapids, into the placid surface. The verge of water, where planting does not occur, should be turfed to 6 in. below the water-line. The slope of the sides immediately below the water-level should be as steep as 1 to 1, so that, in the event of the water falling, there may not be a stretch of muddy foreshore laid bare. Water should never be less than 3 ft. deep, as with less depth than this evaporation takes place with undue rapidity, the bottom becomes heated by the sun's rays, and vegetation is inconveniently stimulated. Single trees of pendulous habit may be planted on the turf, to hang over the water.

The practical work in making a lake must be well directed and thoughtfully planned, or vast labour may be wasted. When the natural soil on the site is not water-holding, or when the level of the water is to come above the line of water-holding stratum, artificial means must be used to prevent unintended escape of the water by percolation. Even on strong clay lands there are generally at least 6 in. of surface-soil, which is more or less porous. Lakes may be made water-holding by use of either cement or "puddle." If the soil be sandy or gravelly, and no clay is available, and above all, if a solid foundation can be found, then concrete may be used. It should be deposited not less than 12 in. thick on the bottom and 18 in. at the sides, and should be composed by mixture of 1 part of hydraulic lime with 4 parts of broken stone and sand, or 1 part of Portland cement with 6 parts of broken stone and sand. This is to be floated with a layer 1 in. in thickness of cement. Before the concrete is laid, the natural bottom must be worked solid and firm. The result is disappointing if the deposit be made on filled-up ground, or on that which is liable to slip, or is otherwise unstable. When puddle is used, many of the same conditions must be created; but there are two ways of using this material. It is made by obtaining clay, then cutting it and cross-cutting it, pouring water on it, and working it in a mortar-mill, or by treading, till it is plastic. The first way of applying the material is this: The ground being excavated to the required depth, allowing for the puddle that is to be deposited, the puddle should be laid not less than 1 ft. thick on the bottom and 18 in. on the sides. As it is laid, it must be well trodden and rammed till the formation is complete and as homogeneous as possible. The second method is more satisfactory, where it is available, and that is when a bed or substratum of water-holding clay underlies the surface at a moderate depth. The whole lake can then be made water-holding by means of puddle-gutters (*see Plan, fig. A*). Such gutters should be based on the water-holding stratum reached by cutting through the overlying soil. From that base must be constructed the puddle-gutter, or contained wall of puddle, to the height of at least 1 ft. above the projected water-level. The gutter should be in no part less than 18 in. thick, and should increase in thickness in the proportion of 1 in 6 as it descends. The plastic clay should be carefully rammed. It may be necessary in some cases to form such a puddle-gutter all round a lake, but in falling ground only at the lower end, so as to prevent water escape. Some lakes may be formed by utilising a natural gully, and throwing a strong dam across its lower end. Then it is necessary to construct a well-made and well-based puddle-gutter in the dam, and the clay wall being amply supported by earth, the surface of the dam may be treated for picturesque effect by planting or other means. The puddle-gutter of a dam is best placed in the centre, and the dam itself should be built up in layers, each one being well rammed and consolidated. It should not have a width less than 3 ft. plus the square of the depth, with slopes on the inside of 2 to 1, and on the outside of $1^1/_2$ to 1. It is well to cover the sides of a lake with gravel, especially during the period of the entry of water, or where there is danger of a wash. In large lakes of considerable depth, and in reservoirs, the inner slope should be pitched to prevent wash, and the water-level should be fully 3 ft. below the top of the embankment. When water enters from a pipe or from any point above the lake-level, the sides below the inlet should be paved: that is, an apron 4 to 6 ft. wide should be laid down the slope. Both the overflow and the emptying pipe, where practicable, should be preferably built in the solid ground. A type of overflow is shown (*see Plan, fig. B*). A grating to prevent the escape of fish or leaves and rubbish through this overflow should be placed at a short distance down the slope, so that it may not be choked by accumulation of leaves, etc.; as they float at the surface of the water, which rises beneath any such accumulation, and so the desired level is maintained. In lakes receiving the drainage of large areas, a storm overflow should be provided; it may consist of a paved duct or channel, formed 3 or 4 in. above the level of the normal overflow. When forming a lake, the operator should bear in mind the practicability of introducing either a hydraulic ram or a turbine, where the useful work it can do may be needed. For example, a stream that fills a 3-in. pipe with a fall of 6 ft. will force the delivery of 2,000 gallons of water daily to a point 100 ft. above the water-level.

A lake may be made in two levels, that are separated by a weir, which should be constructed at some comparatively narrow place, over which a bridge can be thrown; or a cascade can be made at such point; or the bridge may be placed where a short river joins the broader lake area.

In commencing the excavation for a lake, all the soil should first be removed from the area and be stored in a convenient position for use; or be used at once, in position, to form plantations connected with the lake. Such soil, so utilised, has not only its own inherent value, but will give a luxuriant vegetation more quickly than poorer soil, or soil of lesser depth; an advantage, in the special circumstances, often of particular usefulness.

Storage of the water, and maintenance of the proper level, are important considerations. The average evaporation from a water surface amounts to about ⅛ in. per day, or about 30 in. in the year. It is greater from shallow than from deep water, and naturally more in summer than in winter; it is greater from running than from still water.

The mean total annual rainfall in the London district is about 26 in. Heavy rain, falling for 24 hours, yields about 1 in. to the depth; but occasionally rain falls that gives 3 in. of depth in 1 hour. On an average, about three-fifths of the rainfall is available for storage. It will thus be seen that an extraneous water supply is desirable for adjustment of the normal level. A method of approximately ascertaining the quantity of water available from a flowing stream is this: Select some part of it where the cross section is for a given distance constant. Find the area of this cross section in square feet. Prepare a float, and start it from the upper end of the measured space. Count the number of seconds this float occupies in passing the given distance; then multiply the number of square feet by the observed mean velocity in feet per second, and you will obtain the discharge in cubic feet per second. To find the area of a pipe, or a channel-way, or the mean velocity, or the quantity discharged (according to Trautwine) when the other two are given:

$$\text{Area in square feet} = \frac{\text{discharge in cubic feet per second}}{\text{mean velocity in feet per second.}}$$

$$\text{Mean velocity in feet per second} = \frac{\text{discharge in cubic feet per second}}{\text{area in square feet.}}$$

$$\text{Discharge in cubic feet per second} = \text{mean velocity in feet per second} \times \text{area of cross section of pipe or channel in square feet.}$$

These formulæ apply to openings in the sides of vessels, to rivers, and to all other channels, as well as to pipes. According to Hawksley, the formula for the delivery of water in pipes when—

G Number of gallons delivered per hour.
L Length of pipe in yards.
H Head of water in feet.
D Diameter of pipe in inches.

$$D = \tfrac{1}{15}\sqrt[5]{\frac{G^2 L}{H}} \qquad G = \sqrt{\frac{(15\ D)^5\ H}{L}} \qquad V = 48\sqrt{\frac{H\ D}{L}}$$

The theoretical velocity in feet per second is $8{\cdot}025\sqrt{\text{Head of water in feet}}$.

A running spring or a stream delivers about as much water as a pipe having the same area would discharge without extra pressure, and it is worth while to note that a

| | | | | | | | |
|---|---|---|---|---|---|---|---|
| 1 | in. | pipe | discharges | 3 | gallons | per | minute. |
| 2 | ,, | ,, | ,, | 18 | ,, | ,, | ,, |
| 3 | ,, | ,, | ,, | 50 | ,, | ,, | ,, |
| 4 | ,, | ,, | ,, | 112 | ,, | ,, | ,, |
| 6 | ,, | ,, | ,, | 283 | ,, | ,, | ,, |
| 9 | ,, | ,, | ,, | 778 | ,, | ,, | ,, |
| 12 | ,, | ,, | ,, | 1,600 | ,, | ,, | ,, |
| 18 | ,, | ,, | ,, | 4,400 | ,, | ,, | ,, |

In the treatment of a river, everything that conduces to the expression of movement may characterise it, whereas placidity, and the beauty of rest, and finality of boisterous action dignifies a lake. The shores of a lake are circuitous ; the banks of a river are nearly parallel. Where one bank retires, the other, if it does not advance, should at least continue its direction, though the water may be widened to admit of an island being formed, or where there occurs a confluent stream. In nature we generally see at a river-bend some raised ground that has resisted the rush or wearing action of the stream, and so has diverted its course. In such a place the stream is generally wider, and the opposite bank washed low. There should not be bays in the regular line of a river. It is equally an expression of artificial resource, offensively conspicuous, to make the line straight or canal-like. The ending of a river scene should be hidden, like parts of a lake should be. Bridges may, of course, be introduced for utility ; but they may also be contrived to excite the impression of length and extent of the water-way, by conveying an impression of the impossibility of going round or of crossing by other means. It is well so to arrange a bridge that light may be seen beneath it.

# FOUNTAINS.

THE beautiful forms taken by ejected water, rising in a stately column, and falling in glittering spray, have a charm of their own that sometimes is in the nature of a fascination, exercised not only by the water forms, but appealing to another sense by the melody of rippling sound. The arrangement is almost purely artificial, and the resource is most consonant with those parts of the garden where nature is most directly tutored and linked with the direct evidences of art. There is great opportunity for variety in design and application of the beauty of water forms.

Fountain basins differ essentially in their construction from lakes. A regular form or artistic outline must be given to them, dependent, it is true, on their position relatively to the house, or to the surrounding ground, but signalised by its artificial character. The outline is marked by a stone coping, which should possess architectural appropriateness to the neighbouring constructions. If the ground be solid, a brick wall carrying the coping may be built to a suitable depth, as at the Crystal Palace; and the bottom of the basin may be puddled. For small fountains it is usual to form concrete basins, resting against a surrounding wall of brick, on which the coping or kerb is imposed. The inside of such basins should slope from the margin at such an angle that when ice is formed, as it rises and expands, it may find sufficient room and no damage

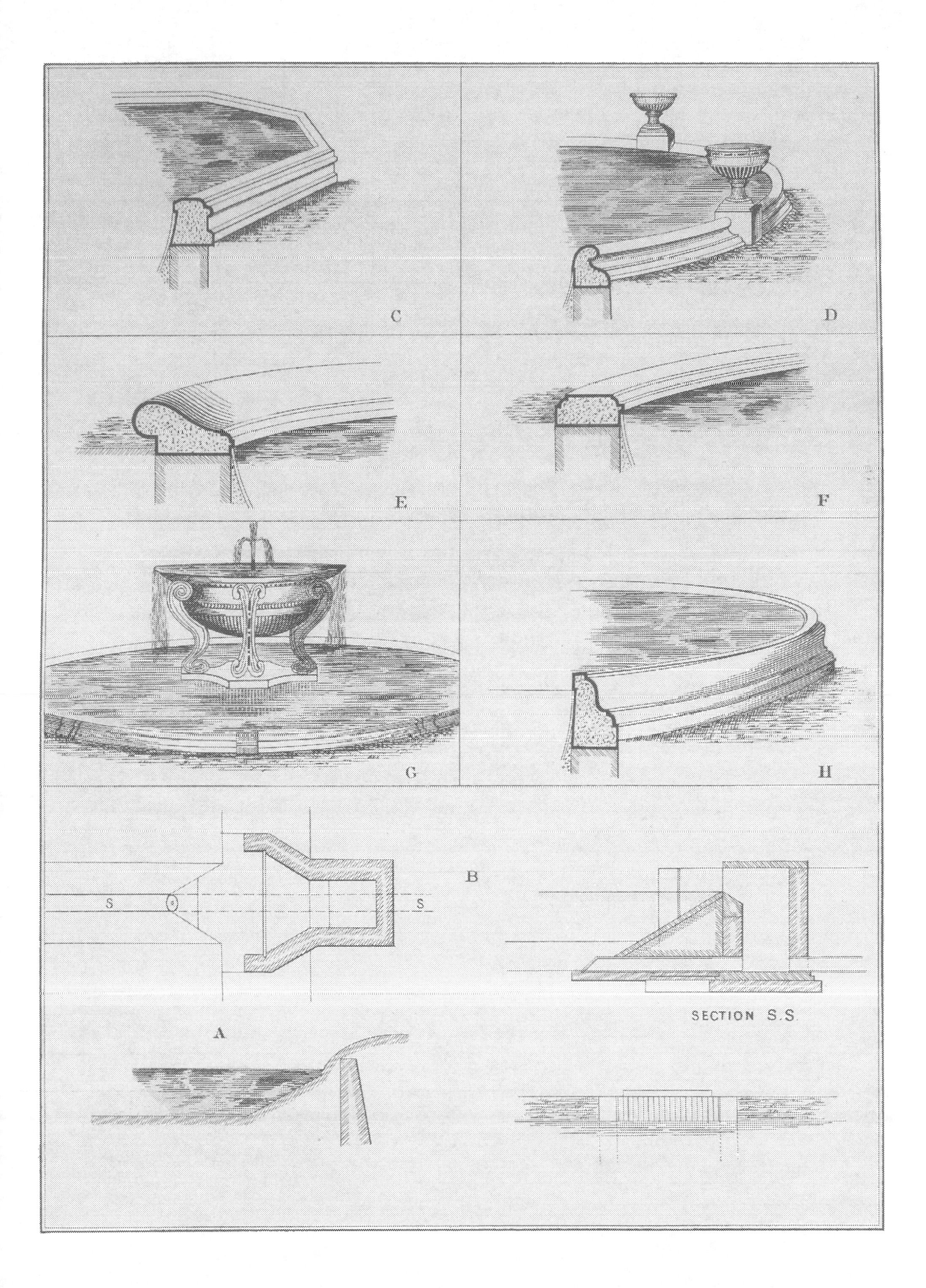
C
D
E
F
G
H
B
S
S
A
SECTION S.S.

ensue. Generally it is desirable to keep the edging not more than 1 ft. above the ground level; but the coping may be made as high as 3 ft., if arranged to be consonant with a particular design. On the parterre in front of the Palace of Count Festetics at Keszthely, in Hungary, the fountain basins are raised 3 ft. above the general level, and from each basin radiate walks to similar stone basins made to contain plants. The effect is good. The drawing (*see Plan, figs. C to H*) presents examples of fountain basins in various styles.

The height to which a jet of water will rise depends on the force derived from the pressure coming from a head of water placed in a higher position, the vertical height of the surface of such supply giving the measure of force to which the jet of water, if directly connected with the supply, will rise; but the height of the jet is also ruled by the size and shape of the nozzle, etc. The height of a jet is not so great as that of the head, owing to the resistance of the air; and the difference between the head and the height to which the jet rises, increases with the absolute height of the jet nearly in the ratio of the square of the head, and diminishes with an increase in the diameter nearly in the inverse ratio to the diameter of the jet. For instance, with 80 ft. head, and with a jet of 1 in. diameter, the loss would be about 10 ft., and the height of the jet 70 ft.; but with a ¼ in. jet the loss would be about 40 ft., and thus the height attained 40 ft. Mr. T. Box is my authority for these statements, and he has worked out the following formula for the height of jets with different heads: $h' = \frac{H^2}{d} \times \cdot0125$, in which H = the head on the jet in feet, $h'$ the difference between the height of head and height of jet, $d$ the diameter of jet in eighths of an inch. Experiments also show that with excessive heads an enormous loss takes place. The quantity of water discharged will vary considerably with the form of nozzle. Assuming this is of good form, it may be found by the following rule (Box): $G = \sqrt{H} \times d^2 \times \cdot24$, in which H = the head of water on the jet in feet, *d* the diameter in eighths of an inch, G the gallons discharged per minute. Thus, with 80 ft. head and ¼ in. jet, the discharge is 8·58 gallons per minute; with ½ in. jet, 34·3; and with 1 in. jet, 137 gallons per minute.

Not only do fountain basins admit of varied formation, but the arrangement of the water display in elevation can be made to assume ornamental forms. The water forms should accord with the art character of the basin. A single jet can only fitly spring from a circular or regular figure; while a combination of jets requires to rise from a surface the outline of which is not only accordant with, but is calculated to set off the water figures. The author condemns the practice of allowing jets to arise from a lake, unless, indeed, that be in form of a basin. A lake should not be made to appear as if it were artificial, whereas a fountain basin is essentially the work of man. Moreover, there should be a distinct difference between an artificial cascade and a natural waterfall. In the former the water should be made in appearance to descend by steps, whereas with the latter, it is a main object to make the water come falling naturally over rock. A sandstone formation lends itself most easily to the production of this effect, as the rockwork can be appropriately stratified in beds of varying thickness and tint. The rock must not look like a wall, but the elevation should be recessed; pockets should be left for ferns. A "fault" may be made in the strata; and it is well in constructing rockwork, to inspect and copy some natural formation of the kind.

# STRUCTURES.

THE word that heads this chapter is intended to refer to a number of art features that are incidentally introduced into the garden or the park, such as Seats, Pavilions, Bridges, etc.; the appropriateness of which, in their design and for their position, is a consideration of importance. And besides these there is the question to be treated of boundary fencing and of divisional fencing, within the limits of the estate—one of the most difficult details of arrangement in the whole subject.

The place for a seat should be chosen most carefully, with full consideration of its aspect, the prospect from its position, and the influence of its surroundings. The object of raising a seat is for rest, and so far as the scene can minister to placidity and retirement, it should be made to do so. If the seat be in a pavilion attached to or near the house, the architectural features of the structure must be consonant with those of the greater building. Such architectural seats will almost always be at no great distance from the house, even when one is placed in an old-fashioned herb garden. In the landscape garden, where positions with purely natural surroundings may be found, the conditions that tempt to rest and delightful contemplation will most frequently be discovered, and there the character of the structure should be, almost without variation, rustic. The first-mentioned seats, temples, or pavilions will be built of wrought stone or brick; the rustic seat, when roofed or sheltered, will be constructed probably of oak or larch posts, with lining of planed wood or hazel strips. The architectural seats, as expressing more art, are directly consonant with artificial treatment in the surrounding features; but the simply rustic seat will be best placed where nature seems to be least touched by art. In the manufacture of such seats it is a mistake—none the less because it is common—to leave the bark on the wood; and the lining of roofed seats is too frequently insufficiently permanent. Forethought is valuable in such cases, and it should be particularly exercised, in regard to the incidence of wind and dust, in every choice of position. Sometimes the structure may be yet more elaborated, and may take a châlet form, as a tea-house, etc., and be worked in pitch pine or other wood; but the foregoing remarks as to position and congruous expression apply with added force. The situation of such a structure may be more prominent. It may well be placed to overlook a formal flower garden, a rose garden, or to terminate a straight walk; or to serve uses near a tennis lawn or other special place.

The design of bridges should be ruled by their apposition to the surrounding work. If a bridge be connected with a straight walk or other formal work, it should be of massive construction with architectural features; if it occur as part of a winding walk, crossing a running stream, it is appropriate that it should be given a picturesque and rustic character. For such a structure peeled oak, larch, or fir can be well used (*see Plate* O). One example given in connection with formal work is that of a bridge over the River Wye, in the public gardens at Buxton, for the main approach from the pavilion to the band-stand, and to the gardens generally; and the second is a rustic bridge over a narrow arm of the lake at Keszthely.

Of mere boundary fencing or paling, there is not much to be

said here. In the division on Planting some observations are made as to the proper treatment of this necessary line, and of the trees that are placed on or near the enclosure. But the intrusion of fencing for useful purposes—such as the exclusion of cattle or deer from the garden — within the boundaries of the estate, requires careful attention. Already, in treating of formation, something has been said as to the making of the boundary between the garden proper and the park. There is a difference in quality of tone between the lawns or trimmed turf of the garden and the pasture of the park or the fields, that is more or less conspicuous as a line of demarcation. The coarser grass is apparent, and the effect is not agreeable. When such a division is close to the eye, that is, not far from the house, it is preferable to make the difference at once marked by a boundary of light iron fencing. But when the landscape, with its features of planting, or distant buildings, cattle or deer, has to be kept in the entire effect of the picture, artifice must be used to save the intrusion of a cutting line, distant though it may be. Though the objects beyond do not belong to the garden proper, and possibly are not under control of the occupier, yet they have been made parts of the picture. When the boundary is sufficiently distant from the sight to make the difference of colour in the turf not so apparent, though a rigid line of fence might be noticed ; or when a place is small in extent, and a line of fencing at right angles to the line of sight would make it appear smaller still, then a hidden fence may be properly used. The ordinary devices that may be employed are: a sunken wall, to the summit of which the ground should be made. The wall should be not less than 4 ft. high, and there should be a level space at its foot of not less than 4 ft. The slope of the ground beyond it should be not less than 2 : 1 ; or a bank can be raised on the inner side of the wall to about a foot above its level. This bank should be turfed, and the eye will then be carried over the wall, hedge, or other intervening object to the prospect beyond (*see figs. I and E*). The line may be broken by raising the ground in places, by groups of planting which may be made to extend outside the division in corresponding positions, and single plants may be employed. Frequently fencing is required round artificial lakes. In such cases approaches to the water for cattle drinking must be provided. As a general rule it is better to make artificial drinking places supplied from the lake, but apart from it, so that cattle may not come to trample and damage the edges. Boat-houses are readily made picturesque features in the landscape. They can be constructed on brick foundations in wrought wood, varnished or painted. The roof should be tiled. In more pretentious structures space can be allowed for a tea-room; but in any case it is well to provide space in the roof for the stowage of sails, sculls, ropes, etc. (*see Plate* R). The entrance to a boat-house can be from the back or at the side. There should be platforms not less than 3 ft. wide on two sides of the interior for embarking or landing; and it is well to make provision for slinging the boats in winter to the main beams.

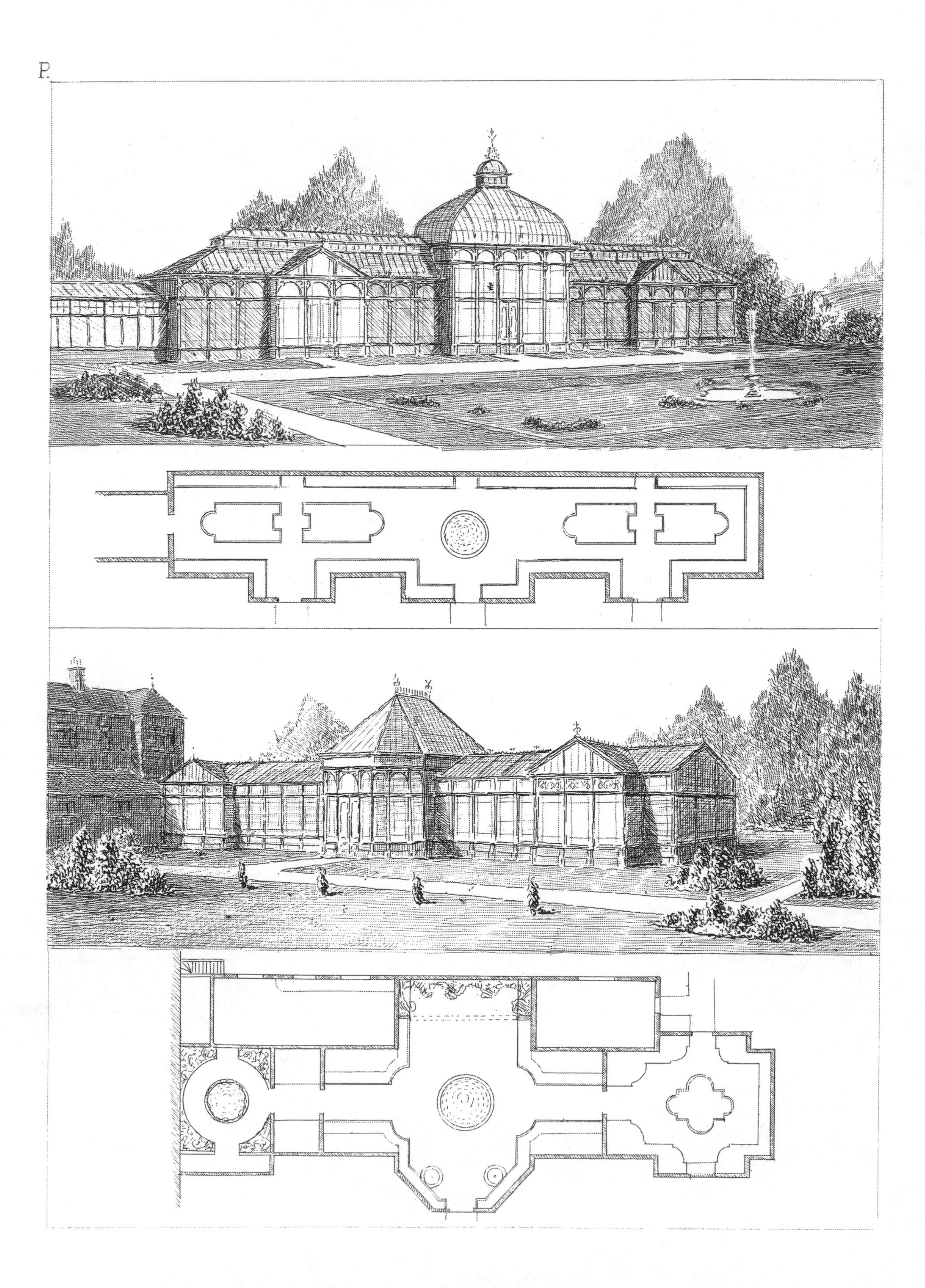
P.

# HOTHOUSES.

Two descriptions of glass-houses are commonly used; one being for the display of plants, the other for their growth. The usual name for the first is the conservatory, while those in the second category are denoted by names signifying the particular use for which each house is designated. Many palms, tree ferns, camellias, etc., usually seen in a conservatory and forming a great attraction, are tall-growing plants, and require a high roof for their display and well-doing; whilst fine-foliaged and rapidly growing plants generally, require the roof to be low, so that they may be near the glass. We have thus two requirements. The large and permanent plants thrive best and look best when planted in beds where they may have an even temperature and moisture to their roots, and where they may have ample space to grow upwards and spread. The foliage and choice flowering plants, not usually permanent, are generally in pots arranged on stages. There results a type of building having a central part with a high roof, and adjacent divisions with as low a roof as is consistent with comfort and good appearance. Conservatories are something more than plant-houses, and must be spacious in order to provide room for walking and lounging. We must then consider the conservatory as an adjunct of the dwelling-house, designed for the showing, under good and pleasant conditions, those plants placed there permanently and those which are being continuously brought there from the growing-houses, such plants being renewed from time to time as required. Seeing, then, that the conservatory is an adjunct of the dwelling-house, and frequently built next to it, it is advisable that the same feeling pervading the design of the house should influence the architectural treatment of the glass structure. If, however, the conservatory is to be anything more than a mere flower-room, it must not be an architectural "feature," having high sides and steep roof, with brick or stone mullions, and an elaborate parapet intended to hide the glass roof—which it frequently does so effectually that it deprives the plants of all light. A conservatory may be treated after this manner only when it is small and unimportant, and when it may be considered as a vestibule or loggia, in which plants are placed. But if the structure is intended to be something more than this, it should, while taking a distinct place in the general composition, and conforming to the architectural character of the house, yet, by its very nature, have a distinctive appearance, and proclaim its object without pretence. Care in the arrangement of the base, the outline of the sashes, the pediment, the entablature, and the mouldings will, without injuring its distinctive features, give the conservatory sufficient connection with the main building to remove all appearance of an excrescence. It is desirable, if possible, to have an antechamber or corridor between a living-room and a conservatory, as the damp air of the latter will, unless modified and controlled by an intermediate space, deleteriously affect the air of the former.

A conservatory is usually constructed two or three steps above the terrace-level, and, as the stages inside are generally lower than those of a greenhouse, we have, outside, a 3 ft. or $3^1/_2$ ft. base of stone or brick, thus leaving, inside, 2 ft. or $2^1/_2$ ft. from floor to stage. The staging may be of stone or terra cotta, as well for the horizontal plant-rest as for the vertical ornamental perforated filling to support the stage and hide the pipes; or it may have a slate plant-rest with iron supports; or it may have wooden laths with half-inch spaces between, on which to stand the plants, with upright wooden supports and lattice work. The height of the cornice or springing of the roof should be about 11 ft., with a transom 1 ft. 6 in. or 2 ft. below this, allowing an upper light to open for ventilation. The pitch of the roof is preferably 30°, as this angle makes the roof half a hexagon, and so facilitates the use of a semicircular rib as a principal, should the design require such construction, and it is also desirable to have an upper part or lantern with lights opening for ventilation. Ventilation should be also given at the bottom by allowing the outside air to pass over the pipes directly through gratings built in the base of the house, or, better still, by taking the outside air through an opening into a channel running inside the pipes, and allowing the air to pass through to the pipes by regular openings. No horizontal sash-bars should be fixed at the height of 5 ft. or $5^1/_2$ ft. from the floor-line, as the line of sight is thereby cut. The floor of a conservatory can be made with marble mosaic pavement, or with tiles of subdued colours. Bright greens and reds should be specially avoided. Pipes below the floor should be covered with a perforated brass grating, or, if expense be much considered, by an iron one, care being taken that the pattern be sufficiently small to prevent the heels of ladies' boots being caught in it. Assuming a stage to be placed next the glass sides, the centre may have a tiered stage; but the space is preferably occupied by a bed surrounded by marble or stone edging. There must be a drain-pipe from the bottom of such a bed. It is very desirable to have space not only for the walk, which should not be less than 4 ft. wide, but also for chairs, tables, statuary, etc. The back wall can be built plainly, or, as at Hitherwood, of alabaster, and treated ornamentally with water pouring from a lion's head into a shell basin; or it can have a border next it, and be wired for the growth of climbing plants; or it can be recessed and lined with tufa, with pockets of soil for ferns, etc.; or space may be arranged for a fountain basin, or for water dripping from the rockwork into a pool. It is well to avoid as much as possible the employment of iron in continuous lengths, and the rafters should be moulded with drips to take away any superfluous moisture. The heating apparatus can be placed under part of the conservatory, or in an adjacent building, or, if near, the heating can be connected with that of the growing-houses. The general details of construction and heating given for hothouses later on, apply also to conservatory work.

Growing-houses are built for utility rather than for show; and therefore it is desirable to construct them with the view of obtaining best results for the well-doing of the plants. A great deal might be written about the best pitch for the roof, or the angle which is made between the roof and a horizontal line drawn from the lowest point of it. The sun's rays naturally strike the glass with greatest intensity when it is placed at right angles to them; but if the sun's rays strike the glass at any angle between 60° and 90°, only about $2^1/_2$ per cent. is lost, so we have practically a range of 30° on each side of the perpendicular. The angle at which the sun's rays strike a surface in the same latitude varies considerably in summer and winter. In summer the angle of the sun's maximum altitude here is 62°, and in the winter it is 15°. In addition, it is well to remember that the difference in the length of a sunlight day in midsummer and in midwinter is 17 hours as against 7 hours. The higher the latitude, the greater should be the pitch of roof; thus for latitude 50°, the pitch would be 36°, and for latitude 60°, 46°. In growing-houses, however, sunlight is not the all-important question. It is desirable that plants should be as near the glass as possible. A pitch that is constructionally good, viz., 26° or 6″ in the foot is a very good

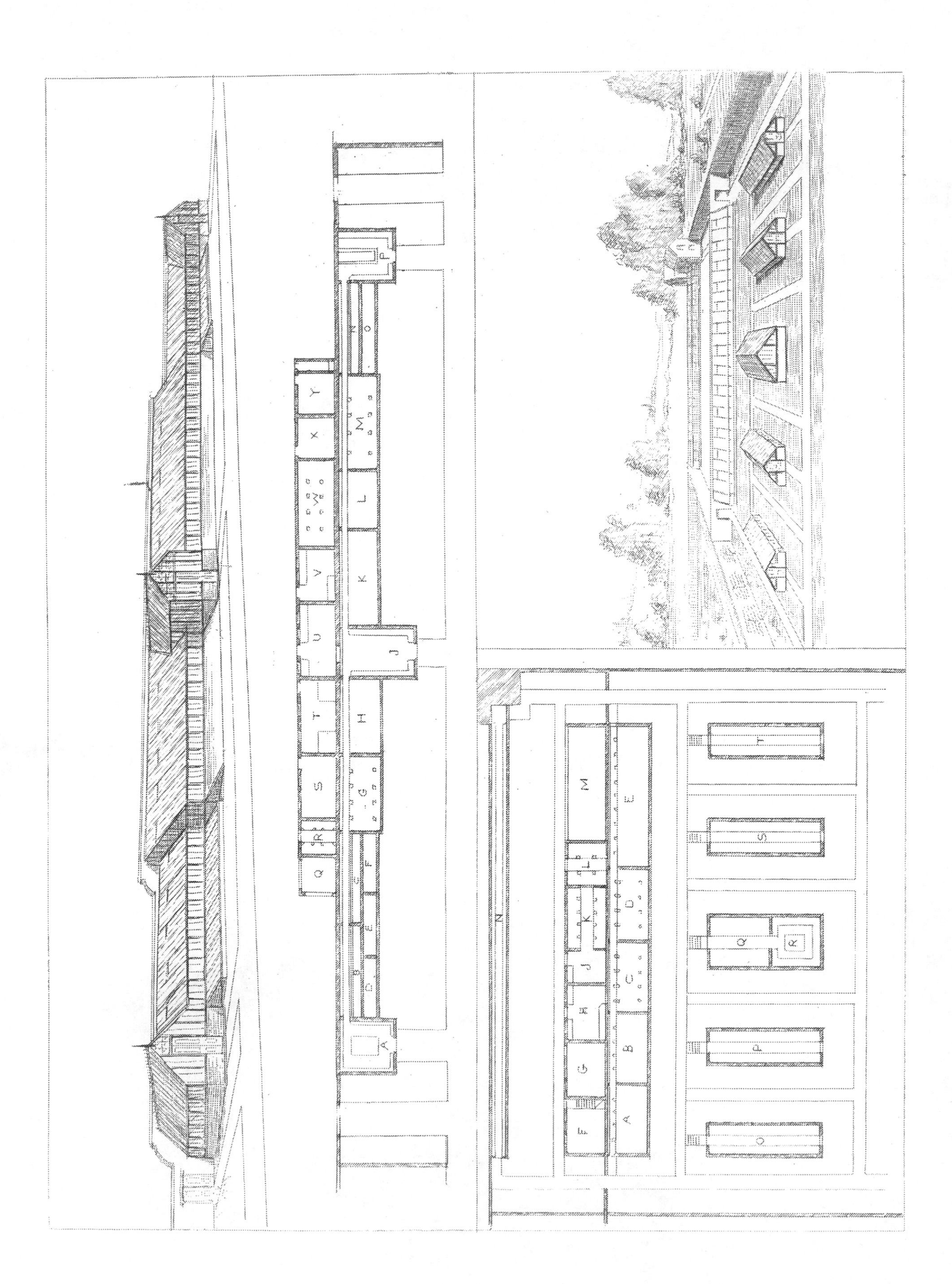

inclination. By making a low pitch, the hot air is more evenly distributed. In houses having a high pitch, the hot air accumulates at the top of the house, and as the tendency of plants is for the sap to flow to the spot where the greatest heat exists, the lower portions of such plants will suffer.

These glass buildings may be divided into two classes, viz., the lean-to houses and the span-roof houses. The former are mostly used for vineries and peach-houses, and are the most economical; the latter are generally used for stoves, forcing-houses, pits, etc. Various examples are given, showing the different combinations alluded to. (*See Plan* Q.) Wood alone is preferable to iron in such constructions, but a combination of iron and wood is advantageous. The plates, sills, mullions, rafters, and posts should be constructed of wood; but purlins, strengthening tie-rods, bars, brackets, etc., may be made of iron.

The houses should be built of best yellow or red deal with oaken sills. This woodwork should be well seasoned, free from defects, and painted two coats before being fixed, and two after. All joints, tenons, mortises, should be primed before fixing. Plate glass is preferable for front lights of a conservatory. The glass for growing-houses should be best 21 oz. English sheet. The front lights and the roof lights should be opened by lever simultaneous opening gear. This gear is constructed by running, through small bearings fixed to each mullion, a rod, to which are fixed at proper intervals the ends of elbow joints; the other ends being attached to the bottom rails of the lights. When the rod is caused to revolve, the whole of the elbow joints are straightened and the sashes thus opened. In conservatory work, sometimes, the vertical sashes are hung from the top, and sometimes at the sides. They are opened by set-opens, which are brass or iron arms, having holes, which are dropped on to vertical pins. Fresh air should be made to enter as near the bottom as possible, and pass out as near the top as may be. In vineries the roots of the vines may be allowed to grow on the inside or the outside of the houses. As a rule, the front walls of vineries and peach-houses are built on arches, so that the roots may spread in every direction. The greatest care must be taken in the formation of a vine border. It is well to lay a pavement, about 4 ft. beneath the surface, of rough material, and from this take pipes, so that the stratum may be perfectly drained; then, on this foundation to fill in a mixture composed of best top spit loam, of $^1/_2$ in. bones, and of soot. The paths inside vineries and peach-houses are best made of perforated iron. In stoves and growing-houses make the footways of concrete floated with cement, or of tiles laid in cement. On falling ground, ranges of hothouses should be built vertically, and stepped when required; though the path inside should be made on the slope, and steps between the houses avoided.

The temperature of hothouses is best attained by employing a low-pressure hot water apparatus. Its operation is based on the fact that water is at its greatest density and minimum volume at 39·2 Fahrenheit. When the temperature is raised above this point, the volume of water increases and the density decreases; so that if we allow a given quantity of hot or cold water to intermingle, the result of difference in the respective specific gravities will be that the superheated water will expand and rise, propelling the colder water, that, having given off its heat, returns to the boiler. And so two columns of water are employed; one is maintained hotter than the other, whilst communication is kept between them, and circulation by displacement is continually going forward. In a boiler there are two openings, the one at the top and the other at the bottom. From the upper opening a pipe is taken, which returns after a certain distance to the opening in the bottom of the boiler. From this point another pipe is led, communicating with a water supply cistern. Assuming now that the boiler and pipes are filled from the supply cistern, which should always be left open, and fire is applied to the boiler, the heated water in it rises, and passing by the opening at the top, through the pipes, becomes cold, and enters the boiler again at the bottom, to be again heated and passed through the pipes. Air or vent tubes are fixed at the highest ends of all pipes, so as to prevent the accumulation of air in them. By hot-water pipes glass-houses are heated, and the heat is diffused by radiation from the pipe surface. Pipes are generally 3 in., 4 in., or 6 in. in diameter. In ordinary greenhouse work 4 in. pipes are used for both mains and heating pipes. For long distances 6 in. pipes are required for greenhouses. The connection from the boiler to the heating pipes may be by 3 in. or even 2 in. pipes. When speaking of heating pipes, 4 in. diameter is generally understood. According to Fawkes, the length of 4 in. pipe required for 1,000 cubic feet of actual atmospheric contents is:

| | |
|---|---|
| Greenhouses, conservatories, etc. . | 35 to 40 ft. |
| Vineries . . . . . . | 45 to 55 „ |
| Plant stoves . . . . . | 55 to 65 „ |
| Forcing-houses, etc. . . . | 60 to 70 „ |

The boiler house and boiler should be placed, in small ranges, at the end, and in very long ranges towards the middle, of the houses. Mains from the boiler should run at a minimum inclination of $^1/_2$ in. per 9 ft.; but the greater difference we have in the temperature between the incoming and outgoing water the shorter may be the column of water to produce circulation, and the higher the column the longer may be the length of pipe. The main should be kept in trenches to prevent radiation, and the pipes for heating the different houses should be taken from the mains at the different points and conducted where required. The heating pipes, once they rise from the mains, should never be allowed to dip. If the pipes must be above the floor-line, and there are doorways, a fresh set of pipes should be arranged between the doorways, otherwise the pipes should be placed below the floor-line in a trench and covered by a grating when necessary. Each house should have its own system of heating controlled by valves from the main. The water should be able to rise to the highest point of the heating pipes as soon as possible, and will thus have less opportunity to cool. Pipes are generally in 9 ft. lengths, with either flange ends, or spigot and socket ends. The first-named are made tight with vulcanised indiarubber joints between the flanges, which are bolted together. For the latter the best joint in underground work is the rust joint, wherein the space between socket and spigot is caulked with rope, and sealed with a mixture of damp iron filings and sal ammoniac. Above the ground the joints may be made tight by indiarubber rings. Boilers vary much, according to different makers. There are very good ones which are most suitable to the class of work for which they are required. The most effective boiler is that which has a large surface horizontally above the flame; vertical surfaces above the flame are reckoned to have half the efficiency of the horizontal surface. Horizontal surfaces below the flame are not valued at all. Every square foot of effective heating surface is reckoned to heat 40 ft. of 4 in. pipe. Coke is the best and most economical heating material. If wood be used, a large grate area is necessary. Clean rain water should always be used for the boilers. It is advisable in a large range to have two boilers, connected or disconnected at will, each one capable of heating the range. It is more economical to have an excess of boiler power and piping than is theoretically required, and to work at a low pressure, than to work a smaller system at a high pressure. Water expands in the boiler about one-thirtieth of its bulk, so the supply cistern should be fixed a slight distance above the level of the highest point of the system and connected with the return pipe near the boiler. The water when cold should always cover the bottom of the cistern 1 in., and as this disappears by evaporation, fresh water should be added. A hundred feet of 4 in. pipes contain 54¹/₃ gallons.

Two ranges of hothouses are shown on Plate Q. The one is at Peverey, near Shrewsbury; the other, at Gisselfeld, Denmark. On reference to the former, it will be noticed that there is a range nearly 300 ft. long, consisting chiefly of lean-to houses placed against the southern side of the north boundary wall of the kitchen garden, with offices, etc., at the back. Beginning at the west end is a projecting span-roof stove, A, with central pit for stove plants, and a stage all round, and hanging shelf from roof. Under the stage are six rows of 4 in. pipes, and under the perforated slate bed of central pit are two rows of 3 in. pipes next the wall. Then follow three pits, B and C, for melons, cucumbers, peas, etc. The heating is given by four rows of 4 in. pipes placed next the pit wall next path, and bottom heat is provided by a flow and return pipe under the pit. The roof of these pits is wired, and a shelf fixed against the back wall. These pits are 9 ft. wide, and in front of them are low frame pits heated by a flow and return 4 in. pipe. The lights of these last named are movable. Then follows an early peach house, G, 16 ft. wide, in which a trellis frame is arranged parallel with the glass and stopping at the path, against which, and the wired back wall, peach trees are planted. The heating consists of two 4 in. pipes between the trellis frame and the outside brickwork, with eight rows of pipes arranged in fours between the trellis frame and the path. Next comes an early vinery, H, 16 ft. wide, having a length of rafters of 18 ft., the heating being provided by eight rows of 4 in. pipes arranged in fours. Then is reached the span-roof plant house, J, forming the centre of the range. There are stages next the wall, but the centre is clear for the storage and display of the larger foliage plants. Next follow a late vinery, K, and a muscat house, L, each 16 ft. wide, each heated by six rows of 4 in. pipes, of which four are 2 ft. from the front and two of them the same distance from the path. The roofs are wired, and shelving is fixed against back wall. A late peach house, M, also 16 ft. wide, succeed, with trellis frame as before described. The heating consists of eight rows of 4 in. pipes, two rows next the outside wall, a group of four rows inside trellis frame, and two rows next the path. We now come to a propagating greenhouse, 9 ft. wide, with stage heated by a flow and return pipe under same, with shelf against back wall, with a frame pit, O, 5 ft. wide, in front, similar to those previously described. P is a span-roof house, with central stepped stage, with tank under the stages next the walls, and hanging shelf. It is heated by four rows of 4 in. pipes under the outside stages. The heating is provided by two low-pressure boilers fixed in boiler house, each capable of heating 3,500 ft. of 4 in. pipe, and so arranged that they can be worked independently or coupled.

There is a main flow and return pipe laid under the perforated iron pathway running the whole length of the houses, from which pipes, controlled by separate valves, are taken to the several divisional houses. All water gathered from the roof is conducted to either the tank fixed under the greenhouse, P, stage, or to an open tank built at back, Q; and pumps connected by lead piping with these tanks are fixed in each house. The front brick walls of the vineries and peach houses are built on arches, so that the roots may penetrate on either side. The front upright lights open and shut by simultaneous opening gear, as also the top ventilating lights, which are 2 ft. 6 in. long, in the roof. At the back are to be seen an open tank, Q, with roof over and enclosed by iron railing, and, as already referred to, a mushroom house, with slate shelves resting on brick supports, and heated by a flow and return pipe, with light and shutter in roof; the fuel store, S; the boiler house, T, necessarily 9 ft. below the ground-line; the potting shed, U, with entrance also from the plant house, J, and fitted with bench, pot rack, bars, etc.; a fitted office and seed room, V, from which is entered the fruit room, W, with stages and grape bottle rests on each side of a central path, having a 4 in. flow and return. There are "hit and miss" ventilators next the ground-line, and a large ventilating shaft in the roof. The walls are built hollow, that is, there is a 2½ in. span between the brickwork, and there is a large light and shutter in roof. Then follow sleeping and mess rooms, X and Y.

In the second example we have a main range of lean-to hothouses 150 ft. long, with five span-roof pits in front, each 40 ft. long, and an orchard house behind, 150 ft. long. The main range embraces an early vinery, A, a muscat vinery, B, an early peach house, C, a late peach house, D, and a late vinery, E, with, at back, a boiler house, F, a fuel store, G, a potting shed, H, an office and seed room, J, a fruit room, K, a mushroom house, L, and an open shed, M; O and P are two stove pits; Q and R are pine stoves; S and T are two propagating greenhouses. The orchard house at the back of the main range consists of a three-quarter span house, 5 ft. 6 in. wide, with the side lights almost vertical. This house is so framed that the lights can be taken off in summer. There is a separate boiler to these orchard houses, owing to the necessities of the climate. A peculiar feature to be noticed is that the roof and vertical lights of the main range and pits have double glazing, as a protection against frost and snow. The roof lights consist of two framed sashes resting on the main rafters, whilst the double arrangement of glazing in the front lights is provided by the additional glazing being fixed on these lights by screws. The outer glazing is removed in summer.

# KITCHEN GARDEN.

THE site of a Kitchen Garden must depend chiefly on convenience, on soil, and on aspect; while the character of the house generally determines the extent of the Kitchen Garden and its propinquity.

For a moderate country house, the area would profitably be two acres enclosed within walls, with space outside the walls and beyond, for the potatoes and coarser vegetables. This enclosed garden should also be easily accessible from the house and at a moderate distance, say between 300 and 500 yds. An ideal site would be on a hillside, slightly sloping towards the south, to the north-east or north-west of the house and intervening stabling; with a wooded hillside above giving shelter from northerly winds. There should be, if possible, good communication with the main road. Water also, preferably that drawn from a river or lake, is a desirable adjunct. If water has to be forced from springs or a stream at a lower level, it is good to have it brought to a service reservoir exposed to the air, with short pipes conducting thence to the houses, and fixed hydrants. It is all-important, especially in clayey soils, that the drainage should be perfect. The depth and distance apart of the drains must depend on the nature of the soil, and if this be hard and impervious, they should be filled up to within 12 or 15 in. of the surface with some porous materials. The subsoil should be improved by trenching, by the addition of manure, lime, or other chemical ingredients; besides which, the top soil may be increased by the addition of soil from outside sources, or if it consist of strong clay, it can be corrected by the addition of sand or slightly burned ballast. A good loam is the best, but it is often convenient to have one part of the garden with stronger soil than the other. The depth of soil should be not less than 2 ft. 6 in. If the subsoil be not good, or if it contain, as often happens, too much oxide of iron, the fruit trees should have a paved space under each, and this must be drained, to prevent roots from penetrating too deeply. When the required soil is brought in from an adjacent meadow, the field should not be stripped, but the surface removed for the depth of one spit in alternate strips, say 2 ft. wide, and the field should be crossed, trenched, and manured.

The form of the garden does not in reality make any difference to the growth of the plants, but it is more easily worked when it takes a quadrilateral form. Much depends on the aspect of the walls, for fruit ripening on them. The sun's rays are generally most powerful between one and two in the afternoon; the consequence is, that a wall with a western is warmer than one with an eastern aspect, though the sun shines on one as long as on the other. To equalise this the walls are frequently placed with a south-easterly aspect, but this must depend on the requirements of the place.

A northern boundary wall is generally used on its southern side for lean-to hothouses, and on its northern for the boiler-house sheds, the fruit-room, mushroom house, etc.; but the wall forming the southern boundary is not so important, for its southern side is outside the garden, and is more or less liable to be shaded by a screen of planting.

Speaking generally, the more southerly the latitude of the place, the more can the line of the walls be turned east of south, so that the southern wall may come at right angles to the sun at 11 a.m.

The higher the latitude of the place, the more must the aspect be turned towards the west, so that the southern wall may be turned at right angles to the sun at one o'clock p.m.

If, however, hothouses are not to be extensively employed, then it becomes important for the sake of ripening fruit that there should be a great length of south wall; and, consequently, a parallelogram is the best form to adopt, with sides as five to three.

An orchard of standard trees should be contiguous to the kitchen garden, and the trees should be planted regularly. It is important that there should be shelter provided when such protection does not naturally exist. Trees make the best screen, and they should be planted on the north, north-west, and north-east sides. In a new plan such a screen might consist of poplar, spruce, sycamore, beech, or pine, as may be best suited to the district. Various forms of kitchen garden are shown in examples (*see Plan*). The borders inside the walls should be from 12 to 18 ft. wide, falling about 6 in. towards the walks. They should be from 2 to 3 ft. deep, well drained, and made with both good top and bottom soil as stated above. If a layer of material be placed under the soil to prevent roots of fruit-trees penetrating too deeply, this layer must be drained, so that by removing the cold stagnant water the soil may be rendered warmer, and capable of receiving air and water warmed by the sun. The main paths in a kitchen garden should be made sufficiently wide to allow the passing of a light cart. The centre walk, leading from the pleasure ground through the kitchen garden to the hothouses, may advantageously be bordered with turf, having immediately beyond a border of herbaceous plants, or of roses or cutting flowers, and beyond that again espalier trees; or it may pass under an arch covered with cordon fruit trees; at any rate, it may be made attractive, and not have vegetables too prominent along its course. A basin forms a pleasant object in the centre of a kitchen garden, with or without a fountain, and it is certainly most useful, as giving a supply of aërated water, besides affording a means for the display of hardy aquatic plants. The walks generally should have a tile edging. If expense be not a great consideration, labour is saved and cleanliness promoted if the walks are made of concrete or asphalte instead of being covered with gravel. The wall forming the northern boundary will often vary in height according to the height or extent of the hothouses placed against it. Speaking generally, the colder the climate the higher should be the walls, for these accumulate heat in proportion to their height. It must be remembered that walls not only retain the solar heat, but they afford shelter, and present a surface on which to train fruit-trees. If there are not hothouses to be considered, the walls on the north side may be 12 to 14 ft. high, while the walls on the east, south, and west sides are 10 ft. high. The south wall may be made the same height or less. The walls can be built equally well of stone or brick. The colour should preferably be of a light shade. For 10 ft. walls, 9 in. brickwork with piers every 10 ft. will suffice; but 14 in. hollow walls are preferable and very little more costly. The walls, especially if built of stone, should be wired on the side against which fruit-trees are to be trained. This surface should, of course, be plane, and the piers, if any, built on the outside of the wall. It should be remembered that the foundations ought to extend at any rate 3 ft. beneath the surface of the ground. Copings can be of stone or brick, or cement, or tiles and bricks, but they should not project more than 2½ or 3 in. beyond the face of the wall. Wide, permanent copings are not so good as temporary copings. Mr. R. Thompson writes in the "Gardener's Assistant" that in summer broad copings prevent the foliage from being moistened by dews, the beneficial effects of which cannot be compensated by artificial watering.

Temporary copings are of great utility, especially during the prevalence of late spring frosts. The heat accumulated in the materials of the wall during the day is abstracted whenever the air is colder than the wall. The cold air coming in contact with the surface of the wall becomes heated, and consequently lighter; it then ascends, and the heat is lost so far as vegetation is concerned. Broad copings obstruct the free ascent of warm air, which then is longer retained where it is wanted—on the surface of the wall. The young shoots of vines may often be seen cut off by frost as far as they have pushed beyond 9 in. coping boards, whilst all the shoots that are under shelter of the boards are safe.

A very good temporary coping is obtained by building in the wall pieces of 1¼ in. galvanised iron tubing, about 6 ft. apart, in which can be placed iron rods projecting 1 ft. beyond the wall, having a pin at the outer end. On this support, in the spring, are placed 12 in. light boards, which are retained there so long as necessary. Netting can also be suspended from these rods if it be required to protect the fruit against frost, birds, or wasps.

| | | Feet. |
|---|---|---|
| The sides of a square containing | 1 acre are | 208·71 |
| " " " | 2 " | 295·16 |
| " " " | 3 " | 361·5 |
| " " " | 4 " | 417·54 |

| | | Feet. | Feet. |
|---|---|---|---|
| The sides in the proportion of 5 to 3 of a parallelogram are, for | 1 acre, | 269·45 × | 161·67 |
| " " " " " | 2 " | 381·03 × | 228·63 |
| " " " " " | 3 " | 466·70 × | 280·02 |
| " " " " " | 4 " | 538·90 × | 323·34 |

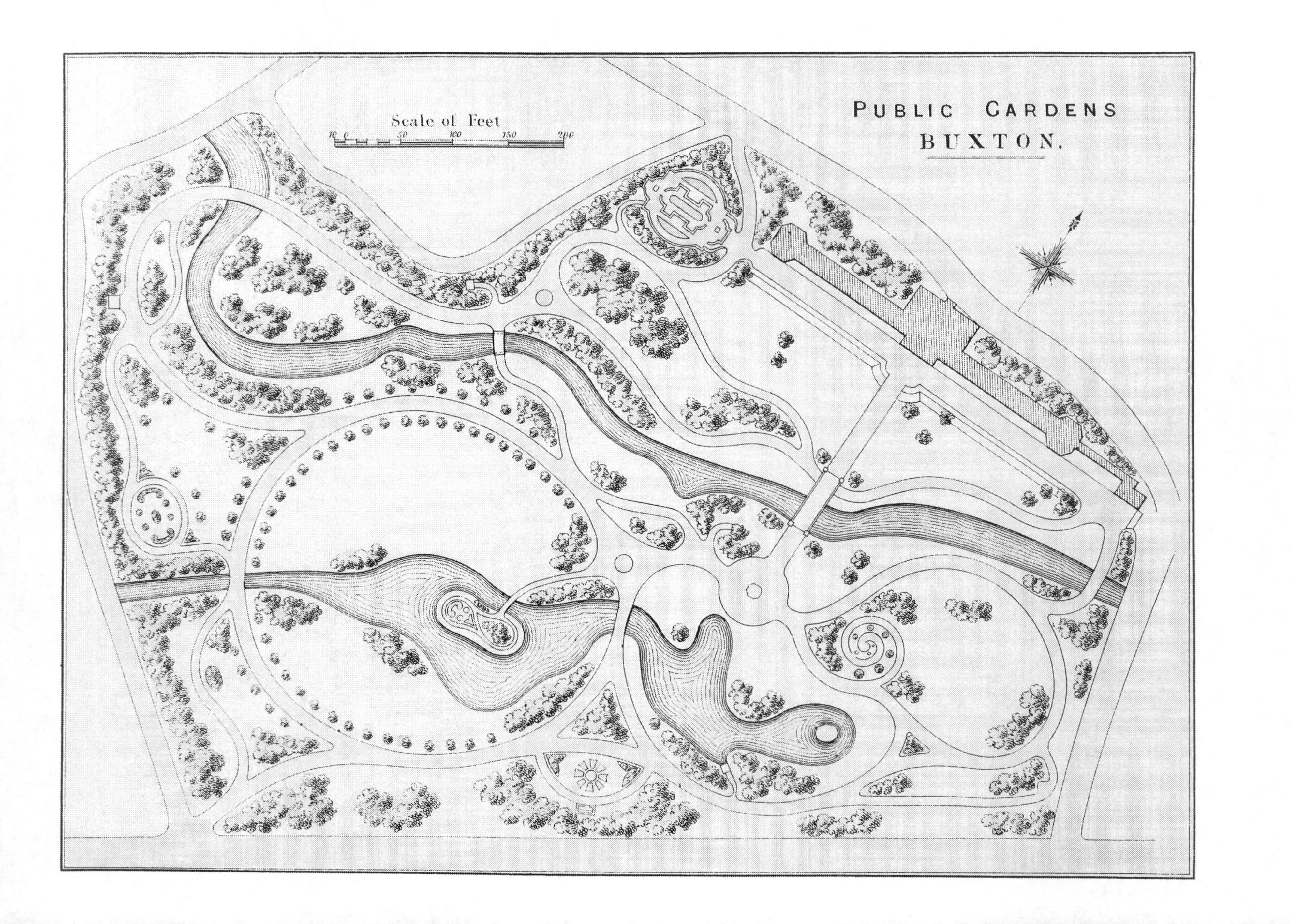
PUBLIC GARDENS
BUXTON.
Scale of Feet
10 0 50 100 150 200

# PUBLIC PARKS AND CEMETERIES.

The observations and rules given for the general treatment of private parks and gardens apply with equal force to the laying out of public parks and recreation grounds, and of cemeteries. There are, however, modifications rendered necessary by the special purpose for which each is intended. Among others is the case where a generous giver, feeling that open spaces are absolutely necessary, presents, ere the builder has invaded the outskirts of a town, some land to be devoted to the public; and the landscape gardener, in dealing with such a property, must appreciate the fact that such land may in a few years prove an intramural oasis; and he must so arrange his laying out and planting that these in the future may bear witness to his forethought. He will, therefore, know that his park may be enclosed by houses; he will arrange for broad gravel spaces; for his turfed ground to be unencumbered with overhead trees; that the plants themselves shall be such as will live amidst smoke and dirt, and he will prepare for what elsewhere are pleasant breezes being converted into biting draughts. In the neighbourhood of towns, too, one is always liable to the ugliness necessitated by modern civilisation, such as the railway embankment, the enforced connection of a park with hospitals, baths, waterworks, etc., or with a large building scheme. At Preston high railway embankments cross the middle of the park and form the western boundary.

Strangers would scarcely notice the railway that crosses the park unless attention be called to the fact by smoke or noise; for the line is so broken by planting and by irregular lines of walk and turf, that the hard, straight course is quite concealed. Rockwork even has been introduced to foster the idea that the towering mass is only one part of an old cliff. The stone of the district is sandstone; so various thicknesses and the dip of these strata are so arranged, that though the lines of rock are constantly broken by turf bays and planting, an observer of nature will find it difficult to recognise that his surroundings are the result of labour, so carefully are the different layers reproduced. At Glossop hospitals and baths formed part of the noble gift to that town, and they had to be introduced and made into integral parts of the whole. In the park some difficulty was created by the existence of a natural ravine with a stream running in the bottom, and completely cutting the land in two. The beautiful and natural parts of this ravine were picked out and made the most of, whilst, in order to convert the parts into a whole, the sides were in places levelled down and the stream covered. A park, too, may be made with the primary object of increasing the value of the surrounding property for building land; and, instead of the desire being apparent that adjacent buildings should be hidden, the object here is rather to furnish an idea that the park is a larger recreation ground belonging to the individual gardens. A place of public resort has generally some objects of particular interest to which prominence can be given and ready access afforded. Such objects are a winter garden, a pavilion, a band-stand, a shelter seat, and in a lesser degree cricket, tennis, and archery grounds, gymnasia, etc. Around the buildings the ground should be treated formally, and near all points of interest where people will congregate broad spaces of drive or walk must be given. In fact, the walks and grassy spaces in public grounds should be wider than those made in private grounds, and the treatment altogether be broader and simpler. In such places, too, people assemble to see and be seen, and it is well to have some special promenade, perhaps emphasized by rows of trees, giving shade, distinctively arranged to serve for a recognised point of rendezvous. Not that the whole place should be considered as a tame plain, visible from all sides, but that the main lines should be drawn with boldness, and a spirit of grandeur asserted. Yet, away from the thronged walks or drives, special charms should be created to tempt visitors with new scenes, and compel them to acknowledge that here is beauty in detail as well as magnificent conception in mass. The area of a cricket ground will in itself give a certain aspect of restful quiet; while broad spaces of turfed ground, with some trees judiciously planted for landscape effect, will emphasize the feeling, and will contrast with the busy world outside as well as will the more secluded winding walks and hidden shrubberies.

A plan is here given of the Public Gardens, at Buxton, as laid out before their pleasantness and appropriateness induced the authorities to extend them, perhaps unmethodically. Part of the site had formerly been used as a garden, and the charm of established trees and water already existed. It will be useful in illustration of some of the conditions already insisted on, to cursorily describe some of the features of these gardens. The principal object was to create an attraction to Buxton, that visitors might be induced to frequent the place, and prolong their stay. These visitors were frequently suffering from rheumatism, and to them the renowned springs were beneficial. The place was, therefore, enclosed, and admission was by a small payment, in return for which good music and other entertainments were provided. A winter garden, about 400 ft. long, was designed, having space for a concert room, promenade, conservatories, etc., all heated by hot water. The main entrance was from the public road, at the eastern end of the winter garden. In front of this, for its whole length, was made a terrace walk, finishing at the western end in a Dutch garden, with clipped yew hedges giving sheltered places for seats, bordered beds arranged for bulbs, flowers, and evergreen foliage plants in their season. Beyond the terrace walk was a grassy slope; on the grass flat between it and the terrace walk vases were set at intervals. From the central hall of the winter garden a main walk, 18 ft. wide, led to the gardens, crossing the River Wye, towards which the ground rapidly fell, by a stone and iron bridge. This walk stopped at a band-stand, from the encircling paths round which other walks deviated right and left. Plate O shows a sketch of the bridge here built, and use was made of the falling ground to take a walk by the river-side, under the side arches of the bridge. The water-level of the river itself was raised by means of the stone dam built to represent a natural cascade. From the space around the band-stand views were obtained within the grounds—on the north, of the river, with its bridges and ground rising towards the winter garden; on the east, of the rose garden, with its spirally ascending walk, thus showing the beds on the slope; on the south, over the lake to the wooded bank beyond; on the west, of the grassy expanse through which runs a small stream, spreading into the lake. A nearly circular walk, with lime-trees planted on each side, is made round the lawn, and walks are taken off at various points, each one more or less concealed, and giving access to some special attraction, either by a rustic bridge crossing the river, giving a peep of a strong spring of water rushing down in a cascade to join the stream; by the formation of a formal garden devoted to various kinds of peat-loving plants; by leading past old trees standing on their grassy mounds and giving shelter; by displaying a flower garden, bright with colour, in front of a covered seat,

STOKE CEMETERY
Consecrated Ground shewn thus
Unconsecrated " "
Roman Catholic " "
Scale of Feet
100
50
0
100
200
300
400
500

tempting to cool repose; or the visitor was tempted to explore and note the grouping, and foliage, and flowers of the many plants displayed, each turn of the walk showing fresh beauty.

In a larger park provision has to be made for driving and riding, and it is advisable for part of the way that the drive and footpath should be made side by side. Space is saved, and persons walking or driving have each an opportunity of meeting. The size and nature of the property, and the purpose for which it will be used, will determine the class of work that is appropriate; but the details already given will suffice for practical work, and I have now rather concerned myself with the expression of such leading ideas as seem called forth by the nature of the undertaking.

And if we provide beautiful places for the living, should we not also prepare for those that have gone before a resting-place, so fitly termed "God's acre," that may coincide with the pure and noble emotions stirring within us, not only at the time we are carrying to their last home the loved ones, but also when we retrace our steps to place on their grave a memorial of our affection? Far too little has been done in this country to alleviate at the saddest moments of our lives the feelings that almost weigh us down. A churchyard or a cemetery can, and should, by the exercise of art, be made as beautiful as possible. The practice of burying in towns is happily being discontinued, and many of the old, dank, sad, neglected, and forsaken churchyards are being turned into new, bright, cheerful, cared-for, and pleasant retreats for the living, where, with the solemn knowledge of the place in which they are assembled, visitors may with ennobled feelings contemplate the various memories that press on them. In a newer cemetery, how much the more should every effort be made to render the place beautiful; and this is right on every ground, both sentimental and economic. If the cemetery be laid out as a park, access must necessarily be given to the chapels, to the different portions set apart for the several creeds, to the mortuary house, etc. The desired area next the walks or drives can be divided off equally well in grave spaces whether the roads are straight or curved; and if, in process of time, that portion of ground taken up at first by groups of planting should be required for burying, the trees and shrubs planted by loving hands at the graves of the departed will serve to replace an effect at first apparent in the general design.

The chapels should be placed in connection with the ground space allotted respectively to each creed, and it is simple to mark the division of such areas by a walk, or, failing that, by mere stones. It is of the utmost importance that there be very efficient drainage, especially of the subsoil. If this is wet, main drains should be made at a depth of 12 ft. from the surface. Except in the first instance, before the ground is occupied, the ordinary surface drainage is of little use, as it becomes disturbed; but the drives and walks should be, therefore, the more efficiently drained. It is also wise in convenient places to fill up the main drains to the surface with porous material, that thus the ground may be kept dry. It is of course necessary that in wet land there be a good outlet for the main drain. The worst of all ground is that where there are alternate and uneven layers of clay and gravel. The best is where a poor, dry, sandy or gravelly soil is found on slightly rising ground. In public cemeteries spaces are generally reserved for the Established, Nonconformist, and Roman Catholic creeds, including special allotments for paupers. The grave spaces are divided into classes according to their nearness to the chapels or the main thoroughfares, and the local authorities impose special regulations for the continuous methodical carrying out of their rules. A reserve ground, with a hothouse, is usually provided near the superintendent's lodge for due maintenance of the embellishment of the place.

The plan of a cemetery made for the Corporation of Stoke-upon-Trent is here given as an example of the manner of laying out now advocated. The approach is from either end of the ground from the public road, and lodges are placed at each entrance. The approach is direct to the two Protestant chapels, which are in this instance joined by a corridor. Sufficient room is allowed here for the circulation of hearses and carriages, and these may be driven to, and pass from, all the principal parts of the cemetery without disturbing the mourners. In this instance a separate chapel for the Roman Catholic service was not required, as the town chapel was not far off. All trees on the property were of course left undisturbed, and they helped to redeem much of that appearance of newness which is always, in such cases, associated with young planting. The first class grave spaces are those nearest to the principal walks; the second class are behind them, and along subsidiary walks; the third class spaces are apportioned in parts next the boundary. The main drains were laid from 12 to 14 ft. deep, at intervals of about 150 ft. The planting was of deciduous and evergreen shrubs, in equal numbers, chiefly of flowering sorts, with bright blossoms, and on raised mounds. The ground surface was undulated so that a play of light and shade might be produced, and the whole work was carried out as though a garden park were being made.

## ECONOMIC TREATMENT OF LAND.

A MOST important division in the work of a Landscape Gardener is the economic treatment of land, whereby the money value of it may be greatly enhanced, as it is made fit for divided residential settlement. The exercise of sound judgment is as important a preliminary, in devoting an estate, or given portions of a land area, to building purposes as to the creation of ornamental ground subserving the purposes of one grand mansion; or to treatment for agricultural uses under the best conditions. Fitness of position is a primal consideration; that is, the relation of the ground to existing routes of traffic, high-roads, or railway communications, or the possible creating and development of such means in the future. The formation of the ground surface is of serious import in the plan, which has to deal not only with general effect, but with particular parts, each part possessing individual features complete in themselves, and, as it were, isolated in their completeness, yet retaining their relation to the whole design of which they are to be component parts. There are statutory obligations to be attended to, involving work not perceptible on the surface, but which, by its excellence and good engineering, greatly raises the value of the estate, and is absolutely necessary to it. But this consideration is paramount: that there should be made with the greatest care a complete plan of the whole area to be treated, showing not only the mode of dealing with it as a whole, but also how it will be treated in detail, as occasion may serve; and that such plan shall be, as it were, a prevailing contract, whereon the working out of each part is based. Upon the excellence of such a plan, and the forethoughtful experience with which it is produced, depends the success of the purpose in view. The money value of an estate may be frequently more than doubled or trebled by such treatment in the course of a term of not many years.

When laying out landed property in the vicinity of towns, for building purposes, there are certain conditions that must rule and have attention: (*a*), Drainage and water supply; (*b*), approach; (*c*), roads; (*d*), proportionate allotment; (*e*), return for money invested; (*f*), common rules; (*g*), general improvement; (*h*), compliance with local regulations.

To treat these in order: (*a*). The question of drainage must necessarily be determined by the site, and its propinquity to an existing drainage system. In the last-named case the connection is simple, provided care be taken in laying out the roads and the sites for the adjacent houses, that the levels are rightly maintained. If no such system exists, the drainage may be made by conducting pipes to a separate sewage farm; by dealing with the refuse matter, and conducting the overflow into a watercourse; or, as a last resource, by means of distribution on the land, or by cesspools. The supply of water also depends on the means available either from an existing source, from an independent provision, or from wells. Where public water supply, or gas supply, exist, the pipes are severally laid by the authorities supplying the same.

(*b*). The approach to any building estate should be made as direct as possible, and every advantage should be taken of existing conditions, such as directing the road into the estate at the foot of a hill, and leading travellers on by an evidently more easy and pleasanter way; by taking advantage of a turn in the public road and arranging the new road into the estate as if this were the continuation of the main road; or, if the road into the estate must be at right angles to the main road, to flatten the corners somewhat towards the principal direction. Should the estate lie somewhat back from the main thoroughfare, it is well to treat the approach to it, if it be through a narrow strip of ground, in a formal manner, and not allow the erection of any small houses which may, perhaps, be appropriate to the available space, but which may, by their smallness, convey an impression of littleness, deteriorating the effect of the whole estate beyond.

(*c*). The roads themselves will vary with every site. Curved roads are preferable to straight ones, because a building estate is generally in the vicinity of a town, and used for residential purposes; therefore it is well to mark the difference between straight streets, and the sweep of a neatly kept road, with its trees spreading overhead, and hiding from view the number of houses, because by the curve the houses are more easily hidden from each other, and a sense of spaciousness is created. Generally, more frontage can thus be obtained. The roads should have a main drain laid down the centre of each, depending for its size on the length of road, the gradient of it, and the number of houses of which it must take the drainage. This drain is usually a pipe, with inlets provided at requisite points, for the general house drainage and for the overflow from the gully holes. The gradient should be easy and continuous. After the requisite excavation has been made to acquire the right gradient-level there should be laid a foundation of stone, brickbats, chalk, or other hard and sound material, at least one foot thick, and on this a layer of gravel or broken stone 4 in. thick. The paths should be raised above the roadway, and have a foundation layer of 9 in. in thickness, and on this 3 in. of gravel. There should also be a granite or other hard stone kerb, and paved

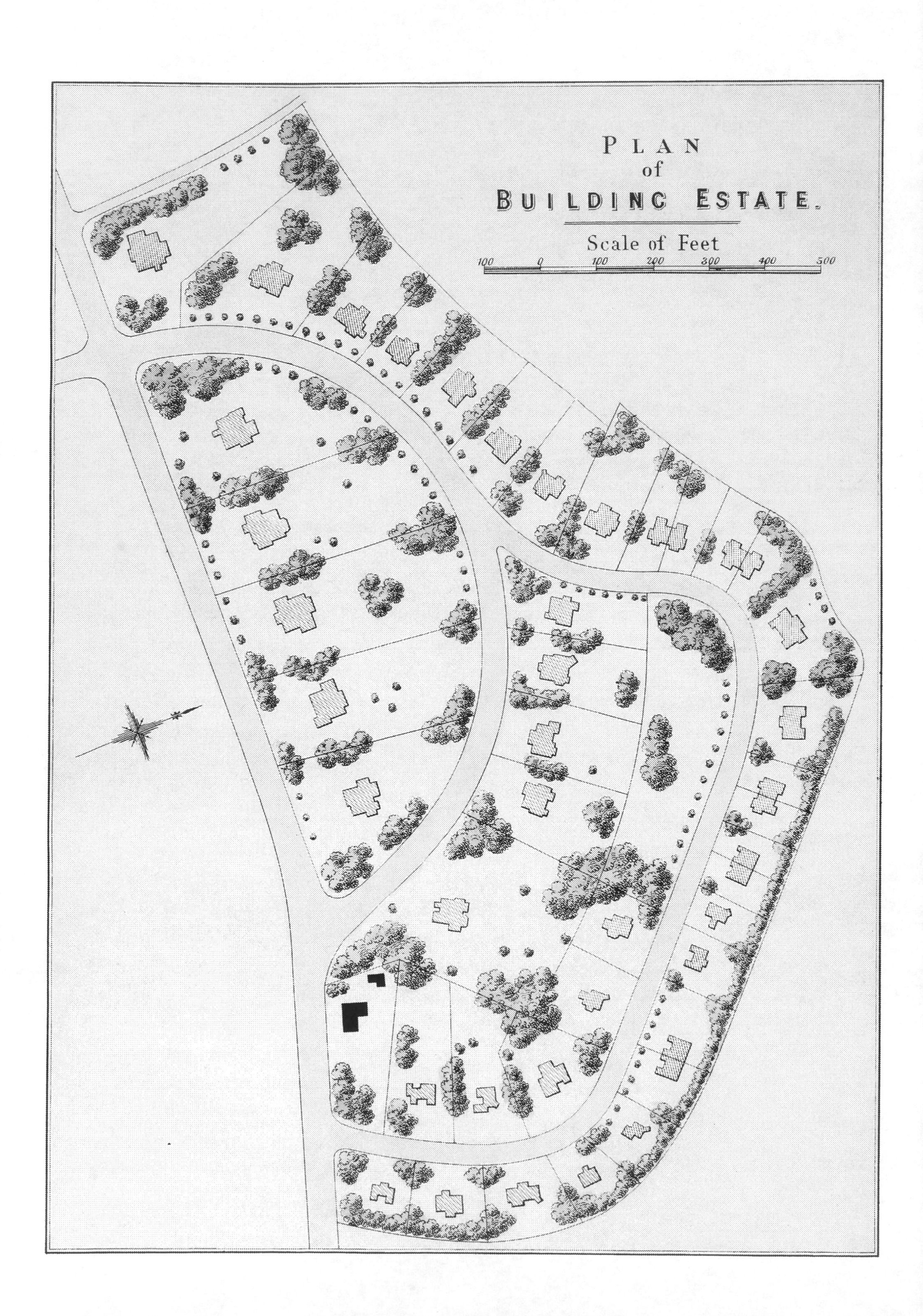

PLAN
of
BUILDING ESTATE.
Scale of Feet
100
0
100
200
300
400
500

channels to conduct the water to the gully holes. It is economical that roads be well made and of good width. If it is considered that an estate will take some time to develop, the centre only of the road may be finished, and a grassy space left on either side, till increased traffic necessitates the completion of it. Trees may be planted on each side by the pathway edge; but it is recommended that these be all of one sort, and of a kind suitable to the district. An ordinary width of road for a moderate estate would be 40 ft., which would allow of a roadway 24 ft. wide and two footpaths each of 8 ft.

(*d*). With regard to the proportionate allotment of the various divisions, no hard and fast rule can be laid down. The frontage and areas given to each lot must vary, as judgment is affected by the nearness of the property to a town, or by the quality of the neighbourhood, and determining the class and value of the house needed. In larger estates which are, most likely, some distance from towns, it is wise to set apart, in the first instance, certain areas of ground, duly divided off, in view of the future development of the property, as occupation land, which may retain its rustic character till the area becomes built over. This enables the landowner to sell smaller portions of land with the houses and, meanwhile, to let on short leases the adjacent ground, as occupation land, till it is required for building on.

(*e*). With regard to the return afforded by the development of land for building, it would be well, perhaps, to state a typical real case—not at all an extreme instance:

A piece of land, 30 acres in extent, one mile distant from a town, has a main road skirting its boundary. The value was £100 per acre, and the land was let for agricultural purposes at £2 10*s*. per acre. The drainage and water supplies were connected with those of the adjacent town. The ground was fairly level, with unimportant views towards the south. The cost of main drainage, of forming the roads, of erecting oak park paling, 5 ft. high, along each side of the roads, of planting the ground for effect amounted to £2,800. There became available for building land 26 acres, after deducting the area of the roads and the ground reserved for common use. The land was sold at an average of £500 per acre, realising £13,000. As the first cost was £3,000, and in addition £2,800 were spent in developing the land, a profit of £7,200 was realised.

In many of our English counties a system of letting the ground on a 99 years' lease prevails, the landowner meanwhile receiving ground rents, and these may generally be regarded as of 25 years' purchase.

(*f*).—The nature and value of the proposed buildings must, of course, differ considerably even on different parts of the same estate, and must be carefully adapted to the respective localities. But when this has been done, it cannot be too strongly laid down that the system for each road or locality should be adhered to rigidly; for only thus will confidence be inspired in builders or private persons erecting houses on the estate. When it is settled that the rules applied to them will be those for their neighbours, and there is no fear of their property being depreciated by the landowner's subsequent dealing, then a security of money value is created. A building line, ruling the distance of structures from the road, should be fixed; and rules should be made for determining the site of stables, so that these may not be a nuisance or interfere with views. Rules should also be made as to cutting down trees, as to the maintenance of the common ground, and the preservation of natural, artistic, characteristic features of the place; also as to the building of shops, public-houses, etc.

(*g*).—It is strongly recommended that plantations should be formed and trees should be planted. These will assuredly make the estate as a whole, and the plots individually, more attractive, besides serving to shut out the objectionable parts of buildings the one from the other. If the planting be judiciously arranged, it will prevent the ungainly effect (but too common on building estates) of each plot being treated wholly irrespective of those adjoining it, whereas, by using the planting as a whole, a pleasing, pictorial scene may be obtained that will add artistically to its present beauty, and, consequently, to its value. It is also advisable on any large building estate to reserve one or more sites for the erection of a church or churches, and schools, and parsonage houses. These may profitably be erected, at any rate in part, as soon as the development of the estate may justify it, or even before it is actually needed. It is a powerful incentive to the taking of building land. A church may be designed so as to be built only in part at first, and enlarged afterwards by the contributions of residents. The church should generally occupy a commanding position, and be accessible from all sides. On large estates it is also advisable to appropriate portions of the land as recreation ground for the common benefit of residents on the estate. In fact, on some of the most successful building estates a central portion has been laid out as a park, to which residents have access. Such privilege forms a powerful incentive to many persons to take land, for it formulates the knowledge that such part cannot be built over, and also allows of smaller plots being sold without taking land for private recreative purposes.

In addition, there are certain statutory regulations relating to the width of roads, drainage, water supply, sanitary condition of houses, etc., which vary more or less in each local government, and must be complied with.

# EXAMPLES OF WORK.

---

## (1) KESZTHELY.

Keszthely is a small town in Hungary, at the southern extremity of the great Platten or Balaton lake. At the western end of the town stood the old family seat of the Counts Festetics. On the western side of the Schloss, a small park had been made about ninety years ago, and many of the trees then planted are now very finely grown to considerable size. Count Tassilo Festetics determined to make this place a palatial home, at the same time maintaining his ancestral castle as far as possible intact. To do this he caused the present structure to be erected, as it were, over, and to contain, the old Schloss. The new Palace was raised from the design of Herr V. Rumpelmeyer, under the immediate direction of Herrn Haas and Paschkis. It then became necessary to consider the artistic treatment of the surrounding land, and the fitting illustration of the grand mansion in its position. To that end I was asked in 1885 to report upon the improvement of the grounds. I found that no view of the distant country could be obtained from any of the principal rooms ; and that on the eastern or true front, the much used, dusty public road passed within twenty yards of the building. Beyond this, the ground sloped gradually down towards the lake, and the whole

Keszthely.

space was occupied by a dilapidated church and small houses of unpleasant appearance, the nearer of which completely shut off all view of the beautiful water, and of the distant volcanic hills beyond. On the northern and southern sides the boundary was very near the Palace, but could not be extended. On the western, or garden front, the old neglected park contained stables and outbuildings. Many fine trees were found on this side, but the ground surface formed a nearly level plain to within twenty yards of the building, down to which it sloped. A considerable innovation and alteration appeared to me to be needed on the northern front. In Hungary it is very desirable to have rooms on the northern and eastern sides on account of the summer heat; in fact, several of the principal rooms in the Palace are so placed. A plan is given showing the general arrangement of the grounds that was adopted near the house.

It will be seen that an imposing arched entrance was built facing the line of the main street of the town, and that the public road was turned at right angles to it, till, with a curve, it was made to reach a distance of 150 yards from the building. Not only was the disagreeable nearness of the public road with its passing traffic removed, but a considerable space of ground was rendered available for private enjoyment. By the removal of a certain number of the intervening buildings, a grand view of the lake and the hills was obtained. Where it leaves the direct line of the public street, the new road was lowered until, by an even gradient of one in twenty-five, a depth of 15 ft. beneath the new levelled garden enclosure was reached; and thus the road, with its passing traffic, was concealed by being sufficiently sunk, while a wall, with planting on the garden side, formed the enclosure; but, as the road became deeper, a balustrade, surmounting a retaining wall, formed the only boundary. From the entrance gates the drive now leads to the chief entrances. The ground inside the newly acquired land was laid out somewhat formally, being intended rather for effect than use. It was made to include a large fountain with jets in form of Prince of Wales' feathers, and two flower gardens were enclosed by yew hedges. The straight walk leading from the principal Palace entrance terminates in a stone pavilion, of similar style to the house. In carrying out this part of the work, several trees, chiefly limes and chestnuts twenty-five to thirty feet high, were moved in the local manner, namely, by digging all round them in the early spring, and leaving such a ball that most of the fibrous root was preserved untouched. On this ball water is poured till the mass becomes frozen. Then the tree and its heavy ball is dragged to the spot it is to occupy. Some of the trees moved a little later by the ordinary English method of digging round and covering the ball with matting, preserving all exposed fibrous roots, and then transporting the tree on a table sliding on planks, gave a better result. By the first method the trees had to be much reduced in size.

On the northern side the area is very restricted, and, as many of the principal guest-rooms look out in that direction, the ground was treated with flowing lines of walks, and with plantations. The surface was undulated, and treated so as to create an appearance of space. Raised mounds were formed to receive the flowering trees and shrubs, and for concealing the boundary. On the south side of the house, where the space to the boundary is also very confined, a formal sunk garden was made, and the beds filled with dwarf *coniferæ.* Beyond this, marble statues alternating with pyramidal golden yews were placed on the flat at foot of a grassy slope—the top of which was planted. Thus the adjacent buildings are completely blocked out.

On this southern side the problems were: (*a*) to provide a

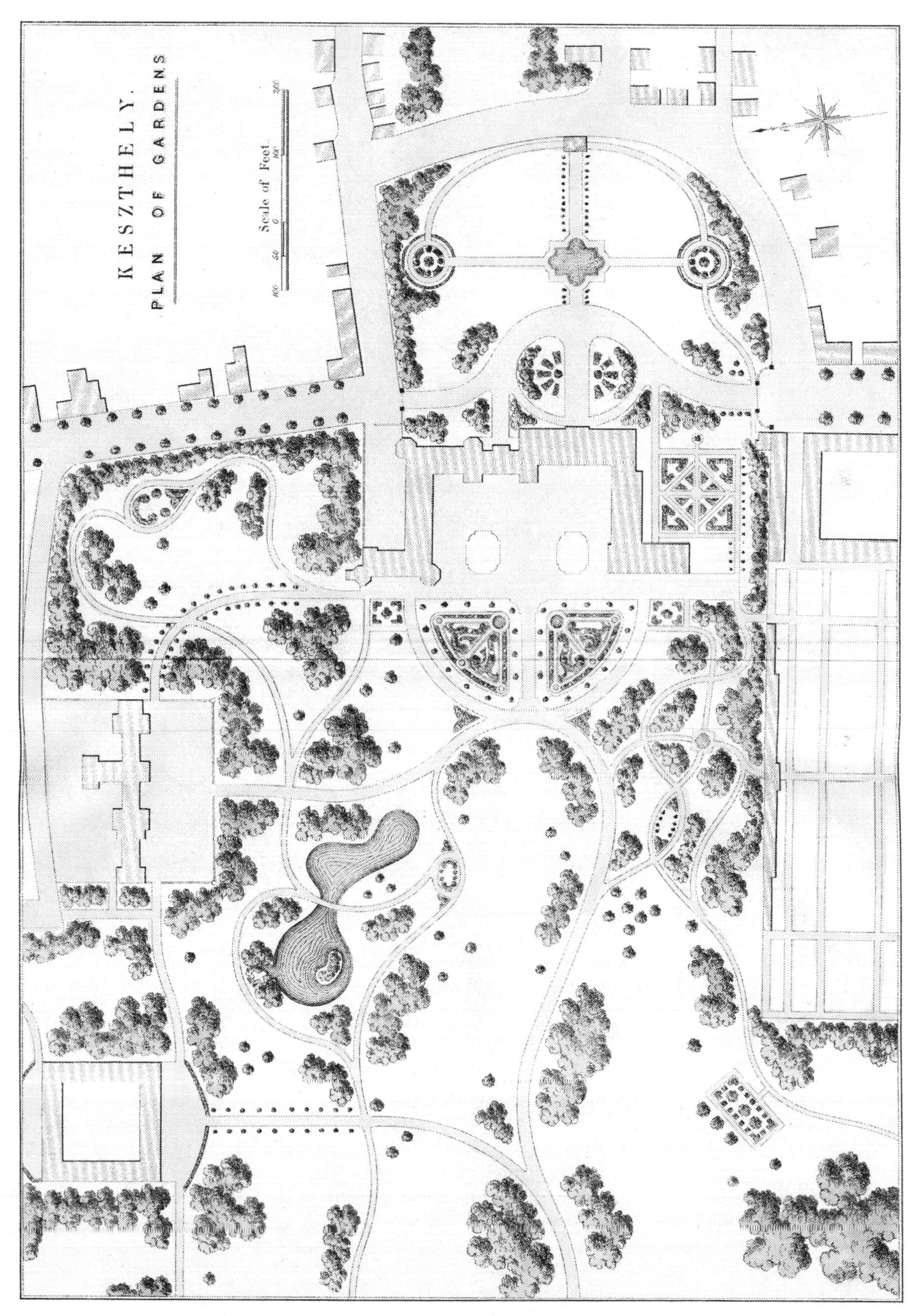


The material originally positioned here is too large for reproduction in this reissue. A PDF can be downloaded from the web address given on page iv of this book, by clicking on 'Resources Available'.

main approach from the western park entrance, with easy access to various points of the house and to the stables; (*b*) to give a base to the building; (*c*) to make a formal garden for the display of fountains, of flowers, and of the sub-tropical vegetation that in Hungary during summer produces such a glorious effect; (*d*) to open out some views, and show something beautiful beyond. (*a*) The first part of the drive from the west, where it enters the park by a double avenue of limes, was naturally determined by that fact; and after crossing a small stream by a bridge, the line curves easily so as to conserve views of remarkable trees, and of vistas in the park, and of distant hills, till a sight of the house is obtained. The drive then leads directly to the formal parterre, of which it becomes part, and so on to the Palace. At the southern end of the parterre there is a direct way to the stables. In addition to the main approach there are several miles of subsidiary drives, each affording glimpses of varied scenery, or serving some special purpose—such as a way to the racing stable, to lodges, kennels, agents' houses, the kitchen garden, etc. (*b*) Much gravelled space was required about the Palace to give sufficient room for several equipages to be present together, and to allow of the many entrances being used simultaneously; so that, in considering the best treatment to be adopted, and to give a base to the Palace, while making apparent the really long extent of its façade, and to provide a line on which this should apparently rest, the question of space had to be taken seriously into account. There is a terrace drive forty feet wide next the building, terminating at its southern limit in a large seat or pavilion, and at its northern extremity narrowing into a drive to the stables. On the western side of this terrace road, the line is fairly continuous; and on the side next the Palace two plots of turf are introduced in the broad gravelled space between the two wings of the Palace, and in line with them, so as to connect these features of the structure. (*c*) The lines of drive next the Palace, the main one from the west, and especially the outline of the south front, helped to determine the shape of the parterre garden. The chief part consists of two quadrants, each having near its centre a fountain basin, of which the stone plinth is $3^1/_2$ ft. above the general level. In each quadrant, walks radiate from the fountain basins towards three other equally raised basins, which are however filled with earth and planted with palms, ferns, and foliage plants. Between these lines of walk the spaces are filled with a geometrical design set out with box edging and coloured gravel walks. On the outer verge of each quadrant, a small rectangular space is made with beds cut in the turf, and planted with dwarf rhododendrons, heaths, azaleas, andromeda, pernettya, etc. The whole of the ground forming the parterre was lowered an average depth of 3 ft., so as to do away with the oppressive feeling that the ground was falling on the Palace. (*a*) On the eastern side beautiful views were opened out over the lake to the distant hills, and on the northern side of the Palace the nearer wooded hills were brought into the picture, and, by means of planting, the sight was directed to certain points. On the east, the town had to be blocked out, and on the west the park was an even plain. From the Palace, the main drive towards the west passed straight through the middle of the parterre, thus creating a line of view. At the end of the parterre, where the drive turns right and left, the line of sight was forced forward by hollowing out the ground for a considerable width, and depositing the earth in raised mounds on either side, which mounds were planted. Trees and plantations were disposed to give length to this main view just out of the direct line of sight; and a clear opening was made for half a mile with an undefined ending in a wood. On either side of the main line of view other openings were created—one towards a small artificial lake, to which the vision was carried over a rustic bridge, to where the water entered, tumbling over a rocky cascade; another over the half-hidden pleasure gardens, past a water temple, into one of the glades in the park. The pleasure gardens were laid out as shown on the plan, beyond the parterre, and special spots were set aside for roses, American plants tennis, shaded walks, and an open-air theatre, of which the walls, divisions, and dressing-rooms were made with hornbeam and privet. Beyond the pleasure gardens, the ground passes insensibly into the enclosed park without any break. Here the trees have been grouped for effect, openings made, and drives formed, to display the beauty of the place. The kitchen garden and hot houses are placed on the south side of the pleasure gardens, and are approached from these, from the park, and also from the outside road. The water for the Palace and for the fountains was obtained by building an engine-house near a spring that runs into the Balaton lake, and by pumping the water to a reservoir built 150 ft. above it in the hills, distant four miles from the Palace. The water thence is delivered by gravitation as required.

## (2) PEVEREY.

In order to practically illustrate the several works incidental to the arrangement of a garden or park, I have chosen to give a general plan and a short description of this place, because there purely agricultural land had to be transformed into a residential estate. The landowner, Sir Offley Wakeman, Bart., determined to build a house on part of his property, and to lay out gardens and a park. The work is not yet actually completed, but, in order to make the description more clear, I am assuming that the whole is finished in accordance with the plan, and I describe the work as it will appear a little later. The River Perry, running from north-east to south-west, divides that part of the property dealt with, and the ground falls gradually towards the river from either side; that part to the north having a slope towards the south, the division south of the river having a slope towards the north. The formation varies somewhat, but generally on the higher ground it consists of about a foot of loam on a clay subsoil, whilst on the lower ground a gravelly or sandy soil is met with. The place is seven miles from Shrewsbury, and two miles from the railway station at Baschurch. Before the work was commenced, the land consisted of a number of fields, opening one to the other, with high hedgerows in which trees, chiefly oak of medium size, were standing. Mr. Aston Webb, the architect, and I fixed the site of the mansion about half-way between the public road, forming the northern boundary of the property, and the river, on a nearly level plateau occurring in the southern incline. The aspect of the site is south by east; very fine views of distant hilly scenery are obtainable towards the south and west, and, in a less degree, towards the east. Dotted lines on the plan show the direction of the principal lines of sight. The house is built of a rich red sandstone, with the principal entrance through an enclosed forecourt on the northern side; the offices are on the eastern side, the living-rooms on the southern side, the conservatory and drawing-rooms on the southern and western sides. The stables were fixed 100 yds. north-east of the mansion in the direction of the home farm, and the kitchen garden 300 yds. west of the mansion on a nearly level piece of ground. Two approaches were desirable, one from Shrewsbury, and the other from Baschurch. For the former, advantage was taken of a sudden bend in the public road to make this apparently lead to Peverey, as shown on the plan, and for the latter, a slight bend in the public road enabled the

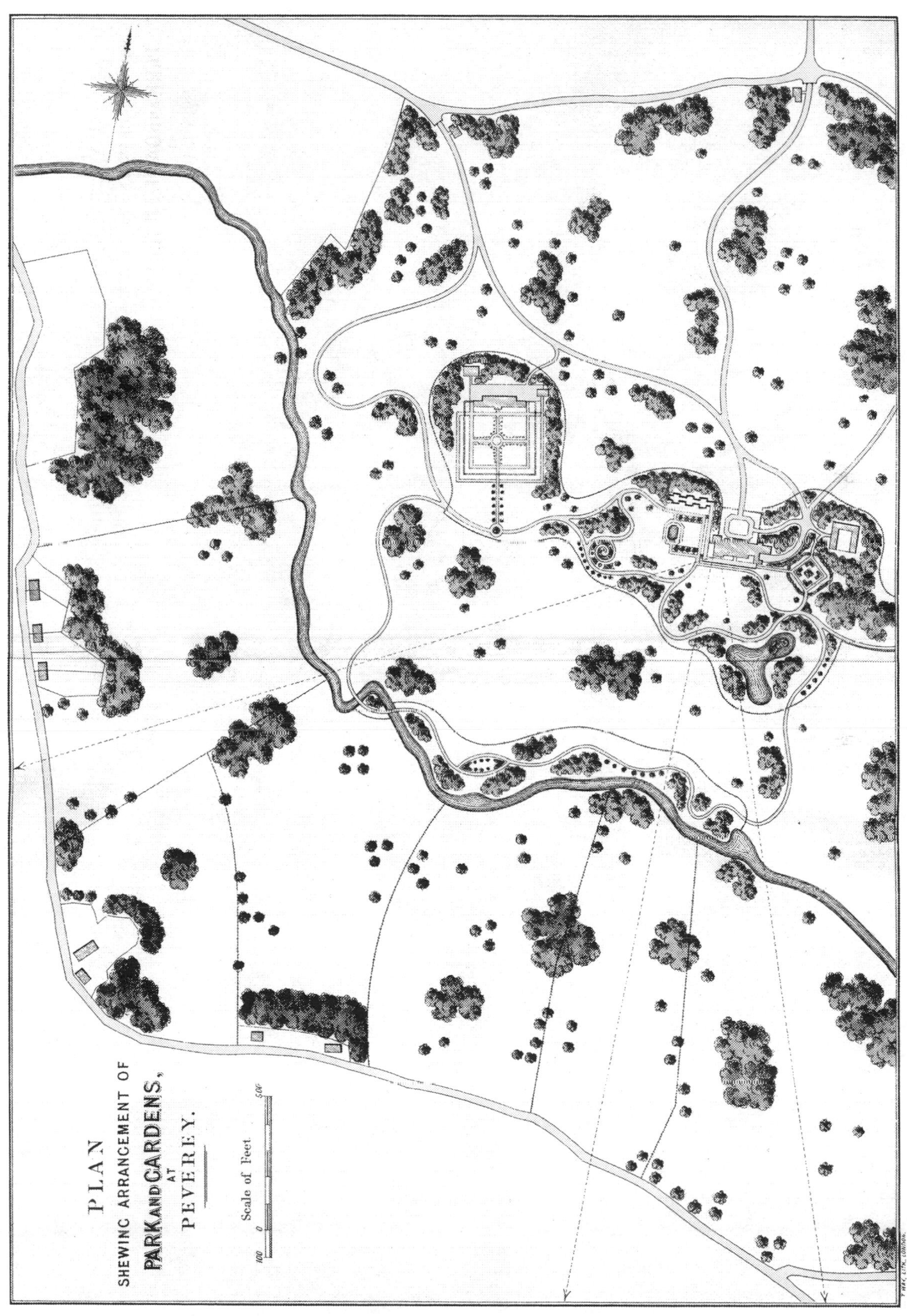


The material originally positioned here is too large for reproduction in this reissue. A PDF can be downloaded from the web address given on page iv of this book, by clicking on 'Resources Available'.

hedge line to be continued with a somewhat sharper curve towards the new entrance. Both drives follow the contour of the land with easy curves and gradients until they meet. From this point the main drive leads directly to the forecourt entrance, and a second drive leads to the stables and offices. Additional communication is also made between the forecourt and stables, and besides that, from the stables a back service road leads past the home farm, laundry, etc., to the public road, which it enters at a point just beyond the Shrewsbury entrance lodge. The ground north of the river was somewhat bare of trees, so large masses of planting, besides single trees, have been introduced for effect. A large belt was made round the exposed sides of the new kitchen garden, to provide shelter and to hide it from the mansion; another continuous belt was planted on raised ground to the north of the pleasure gardens, to screen those from the north wind and from the drive. On the southern side of the house a broad terrace walk was formed, supported by a stone wall 6 ft. high, surmounted by a balustrade, with projecting bastions, and recessed on its face to give space for seats available from a walk formed on the lower level. This work was designed by Mr. Aston Webb. At the eastern end, the terrace walk is narrowed in line with the projecting wing of the house, and is continued to a diamond-shaped garden, enclosed and divided up by yew hedges, with secluded walks and spaces for seats, herbaceous borders, etc. From the northern corner of it, a walk leads to the stables, and from the southern corner another leads to the pleasure gardens. On the western side of the house is a shorter and narrower terrace walk, beyond which is a grassy slope, at the foot of which is a parterre containing flower beds in geometrical design, and a fountain basin 60 ft. by 40 ft. There are steps leading from the western end of the main terrace walk to this lower ground, and also from a walk leading to it from the forecourt. At the northern end of the western terrace is a turfed *allée* enclosed by a holly hedge, with a group of statuary at the lower end, and recessed spaces for seats at regular intervals in the sides. A small lake, wherein bright waterfowl may disport themselves among several kinds of aquatic plants, has been formed in the south-eastern part of the pleasure gardens. The end of the lake is left unplanted, and the view is thus extended over the park past an old red-tiled mill to the distant hills. Around the lake, and below the terrace generally, the area has been treated in the natural manner. The ground has been undulated, grassy valleys have been formed between raised mounds planted with varied flowering trees and shrubs, or with single trees, so that a play of light and shade is secured. The walks have been undulated in plan as well as level, and the different parts of the garden are hidden and separated to such an extent that the whole cannot be seen at once, and the former plain field-look has been entirely removed. At the same time a feeling of rest has been made to pervade the whole treatment which is consonant with the stately building, the home of an old English family. On the south-western side a principal attraction is the rose garden, in which different kinds of roses are planted in separate masses. It is made in a spiral form, with the walk gradually descending towards the centre, so that the beds of roses are all on a slight slope facing the spectator. This position is much shut off by planting, and quite hidden from a contiguous walk that leads more directly towards a covered seat. This is placed at the extremity of the principal walk made from this point, through the centre of the kitchen garden, to the middle of the range of hothouses. A row of standard hollies has been planted on each side of the walk till the kitchen garden is reached, when a line of espalier fencing, along which horizontally trained trees have been planted, borders the walk. In the centre of the kitchen garden is an old dwarfed oak, round which the walk circles, and other walks are made at right angles, having at their ends small water basins to which the overflow from the hothouse tanks has been conducted. The walks forming the principal squares of the kitchen garden have pyramidal fruit-trees planted on their inner sides; the walls are planted with suitable trained trees. From the before-mentioned covered seat a walk is prolonged outside the pleasure gardens across the park, and, dividing, one branch turns northward, to give a view of the country and river not obtainable elsewhere, and eventually joins the drive, whilst the other is continued to the river-side walk. Only small peeps of the river were obtainable from the mansion, so in order to bring this water view into the picture, the river has been widened in places, and the level raised where this could be readily done and could be most visible from the house and grounds. A walk has been made by the river-side, which, starting from the eastern end of the pleasure gardens, crosses the park and then skirts the water's edge. In this secluded part, which has been fenced in, places have been reserved for a collection of bog plants, for a Rhododendron garden, for a bulb garden, and for a collection of willows, etc. The walk, after crossing a bend of the river by two rustic bridges, eventually joins the route already referred to coming from the west end of the pleasure gardens.

In the grounds, lying to the south of the river, and facing the mansion, many trees in hedgerows running at right angles to the eye had to be removed, in order to break up these lines and also to open out views. Of those trees only a few could be transplanted. Groups of planting have been introduced for general effect, to lead the eye in the desired directions, and to give distance by affording so many objects to count just outside the line of sight; to give variety in colour, to give masses of wood, in order to relieve the monotony of the grass land, to provide cover, and to shut out some adjacent land on which, possibly, objectionable buildings may be raised. All ground was doubly trenched previous to being planted, and the whole land drained. The lines of fencing for dividing off the park into different holdings were arranged so as to be as little seen as possible, and therefore are chiefly set out in a direction running from the eye. The hedges forming the boundaries of the public road on the southern side of the park were in places removed, so that the line of sight might not be interrupted, and new fencing was substituted. All fencing was of the usual continuous bar fencing. The height of the wall on the northern side of the kitchen garden was determined by that of the hothouses built against it; the height on the eastern and western sides is 9 ft. 3 in. to the under side of the movable coping; the height of the southern boundary wall is 6 ft. The water supply is provided by an engine pumping river-water to a reservoir near the home farm, and for the garden supply an open reservoir has been made from which water gravitates to given points. The lake is supplied by waste water from this, and from the house and garden clear water drainage.

CHARLES DICKENS AND EVANS, CRYSTAL PALACE PRESS.

Printed in the United States
By Bookmasters